The Open University

Mathematics Foundation Course Unit 19

RELATIONS

Prepared by the Mathematics Foundation Course Team

Correspondence Text 19

The Open University Press

Open University courses provide a method of study for independent learners through an integrated teaching system including textual material, radio and television programmes and short residential courses. This text is one of a series that make up the correspondence element of the Mathematics Foundation Course.

The Open University's courses represent a new system of university level education. Much of the teaching material is still in a developmental stage. Courses and course materials are, therefore, kept continually under revision. It is intended to issue regular up-dating notes as and when the need arises, and new editions will be brought out when necessary.

Further information on Open University courses may be obtained from The Admissions Office, The Open University, P.O. Box 48, Bletchley, Buckinghamshire.

The Open University Press
Walton Hall, Bletchley, Bucks

First Published 1971

Printed in Great Britain by
J W Arrowsmith Ltd, Bristol 3

SBN 335 01018 0

Contents

Objectives

The general aim of this unit is to introduce two important types of relation: the *equivalence relation* and the *order relation*. After working through this unit you should be able to:

(i) distinguish between a relation, a function, a mapping and an operation;
(ii) determine whether a given relation has the reflexive, symmetric, anti-symmetric or transitive properties;
(iii) determine whether a relation with given properties is an equivalence relation or an order relation;
(iv) determine the equivalence classes of a given equivalence relation and construct the natural mapping;
(v) determine whether a given operation on a set is compatible with the natural mapping;
(vi) given a set S with an order relation defined on it, give examples of upper and lower bounds for a given subset of S, and determine the greatest lower bound and the least upper bound, if these exist.

Note

Before working through this correspondence text, make sure you have read the general introduction to the mathematics course in the Study Guide, as this explains the philosophy underlying the whole course. You should also be familiar with the section which explains how a text is constructed and the meanings attached to the stars and other symbols in the margin, as this will help you to find your way through the text.

Structural Diagram

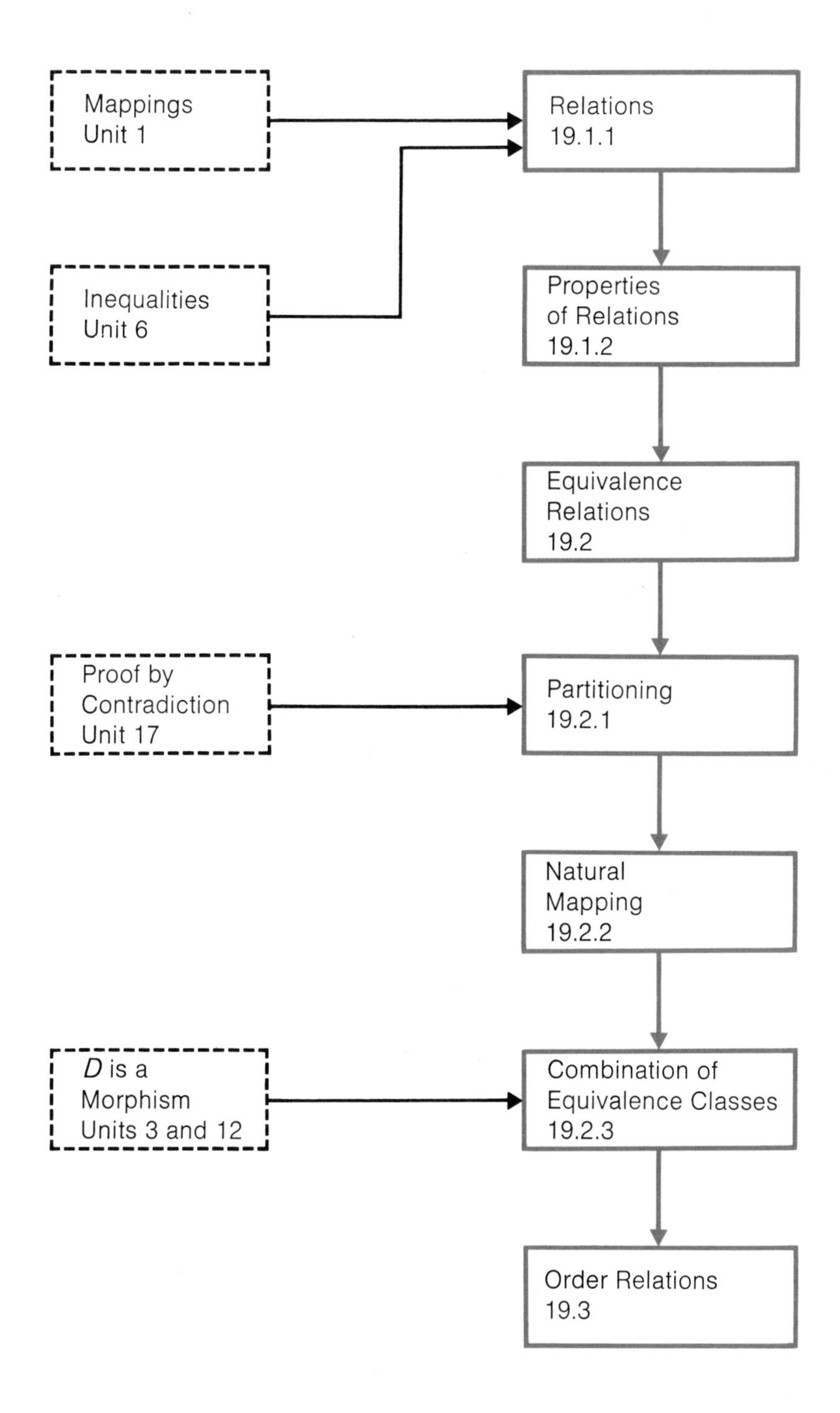

Glossary

Page

Terms which are defined in this glossary are printed in CAPITALS.

Term	Definition	Page
ANTI-SYMMETRIC RELATION	An ANTI-SYMMETRIC RELATION on a set A is a RELATION ρ having the property: if $x \rho y$ and $y \rho x$, then $x = y$ $(x, y \in A)$.	17
COMPATIBLE	An EQUIVALENCE RELATION ρ and a binary operation $\circ$, both defined on a set A, are COMPATIBLE if and only if $(x_1 \circ y_1) \rho (x_2 \circ y_2)$ whenever $x_1 \rho x_2$ and $y_1 \rho y_2$ $(x_1, y_1, x_2, y_2 \in A)$.	28
EQUIVALENCE CLASS	An EQUIVALENCE CLASS of a set A is a subset belonging to a PARTITION of A by an EQUIVALENCE RELATION.	24
EQUIVALENCE RELATION	An EQUIVALENCE RELATION is a RELATION which is REFLEXIVE, SYMMETRIC and TRANSITIVE.	20
GREATEST LOWER BOUND (g.l.b.)	The GREATEST LOWER BOUND (g.l.b.) of a subset S of an ORDERED SET P is the LOWER BOUND l_g, such that $l_g \in P$ and $l \preccurlyeq l_g$ for every lower bound l of S; l_g may not exist.	35
HASSE DIAGRAM	A HASSE DIAGRAM is a diagram illustrating an ORDER RELATION.	33
LATTICE	A LATTICE is an ORDERED SET in which a subset composed of any two elements has both a LEAST UPPER BOUND and a GREATEST LOWER BOUND.	38
LEAST UPPER BOUND (l.u.b.)	The LEAST UPPER BOUND (l.u.b.) of a subset S of an ORDERED SET P is the UPPER BOUND, u_g, such that $u_g \in P$ and $u_g \preccurlyeq u$ for every upper bound u of S; u_g may not exist.	36
LOWER BOUND	A LOWER BOUND of a subset S of an ORDERED SET P is any element $l \in P$ such that $l \preccurlyeq a$ for *all* elements $a \in S$.	35
NATURAL EQUIVALENCE RELATION	The NATURAL EQUIVALENCE RELATION on a set A is the RELATION defined by: $x \rho y$ if and only if x and y have the same image under a given function.	26
NATURAL MAPPING	The NATURAL MAPPING is the mapping of a set A under which elements of A are mapped to their respective EQUIVALENCE CLASSES under a given EQUIVALENCE RELATION on A.	26
ORDERED SET	An ORDERED SET is a set with an ORDER RELATION defined on it.	33
ORDER RELATION	See PARTIAL ORDERING RELATION and TOTAL ORDERING RELATION.	
PARTIAL ORDERING RELATION	A PARTIAL ORDERING RELATION is a RELATION which is REFLEXIVE, ANTI-SYMMETRIC and TRANSITIVE.	33
PARTITION	A PARTITION is a separation of the elements of a set into subsets such that each element is in one and only one subset.	23
QUOTIENT SET	The QUOTIENT SET of a set with an EQUIVALENCE RELATION defined on it is the set of all EQUIVALENCE CLASSES.	26

		Page
REFLEXIVE RELATION	A REFLEXIVE RELATION on a set A is a RELATION ρ having the property: $x \rho x$ for each $x \in A$.	13
RELATION	A RELATION consists of two sets A and B, together with a relationship connecting any element of A with any element of B in a prescribed order, which either holds or does not hold for any choice of the two elements.	9
SOLUTION SET	The SOLUTION SET of a given RELATION connecting sets A and B is the set of ordered pairs $(x, y) \in A \times B$ for which the relationship holds.	8
SYMMETRIC RELATION	A SYMMETRIC RELATION on a set A is a RELATION ρ having the property: $y \rho x$ whenever $x \rho y$ $(x, y \in A)$.	15
TOTAL ORDERING RELATION	A TOTAL ORDERING RELATION on a set A is a PARTIAL ORDERING RELATION ρ on A such that: for all $x, y \in A$, $x \neq y$, either $x \rho y$ or $y \rho x$.	33
TRANSITIVE RELATION	A TRANSITIVE RELATION on a set A is a RELATION ρ having the property: $x \rho z$ whenever $x \rho y$ and $y \rho z$ $(x, y, z \in A)$.	18
UPPER BOUND	An UPPER BOUND of a subset S of an ORDERED SET P is any element $u \in P$ such that $a \preccurlyeq u$ for all elements $a \in S$.	35

Notation

Page

The symbols are presented in the order in which they appear in the text.

Symbol	Meaning	Page
$A \nsubseteq B$	A is not a subset of the set B.	5
$a \rho b$	a is related to b.	7
$a \not\rho b$	a is not related to b.	7
$\forall_x$	The universal quantifier (read as "for *all* x").	13
$\Rightarrow$	The symbol for implication.	17
$\wedge$	The symbol for conjunction.	18
a	The proposition **a**.	18
$\varnothing$	The empty set.	23
A/ρ	The quotient set (read as "A by ρ").	26
n	The natural mapping $n: A \longmapsto A/\rho$.	26
$[x]$	The equivalence class containing x.	28
f'	The derived function of the function f.	30
D	The differentiation operator.	30
$\prec$, $\preccurlyeq$	The symbols of ordering.	33
g.l.b., l_g	The greatest lower bound.	35
l.u.b., u_g	The least upper bound.	36

Bibliography

S. Selby and L. Sweet, *Sets, Relations, Functions*, (McGraw-Hill, 1963).

This book provides a general introduction to sets and relations. Chapter 3 is the most directly relevant section of the book for this unit, but almost all the contents will provide useful material as background to the general concepts of set, relations on a set, and mappings from one set to another.

T. Donnellan, *Lattice Theory*, (Pergamon, 1968).

The first chapter deals with sets and relations with particular reference to equivalence relations. The remainder of the book is concerned with lattices, starting from the concept of a relation of partial order.

D. E. Mansfield and M. Bruckheimer, *Background to Set and Group Theory*, (Chatto and Windus, 1966).

There is quite a full treatment of equivalence relations in this book and, unlike the first two books recommended, it does include a discussion of compatibility.

19.0 INTRODUCTION

From a very early age, one implicitly realizes the need for *sorting* and *ordering* things in various ways. A young child, when playing with a collection of objects (bricks, rings, counters), will often sort them into heaps, perhaps according to colour, or pile them into some sort of pyramid, thus ordering them according to size. In later childhood, many children collect stamps, sorting them according to their country of origin, and sticking them into an album where these countries are ordered alphabetically.

At school, the way in which one is sorted and ordered according to marks gained in examinations becomes very important. You can probably think of many situations of this kind. A librarian, for instance, sorts his books into classes according to their subject matter, and then orders the books within any one class alphabetically according to authors. The classes are usually each allocated a decimal number which depends upon the actual system of classification adopted, and the classes are then ordered numerically and located in the various bookcases according to this numerical order.

Because sorting and ordering are such common everyday phenomena, mathematicians ask themselves whether such processes have intrinsic features which can be abstracted in order that mathematical models of the processes can be made. To discover the answer to this question it is first necessary to describe sorting and ordering situations in mathematical language. Before doing this, however, let us take a closer look at two of the particular examples which we have already mentioned.

Let us suppose that a young boy named Peter has a collection of different coloured cubes of various sizes. He may enjoy himself for a while sorting these cubes according to colour, and perhaps will finish up with four piles of cubes:

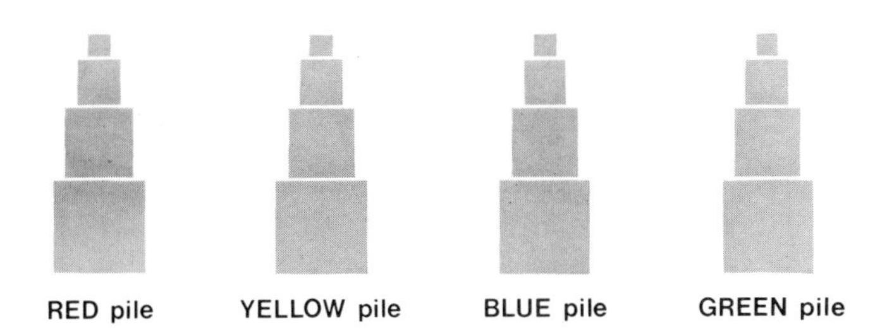

Now let us suppose that he is visited by his friend Paul, who promptly moves the piles around so that he finishes up with:

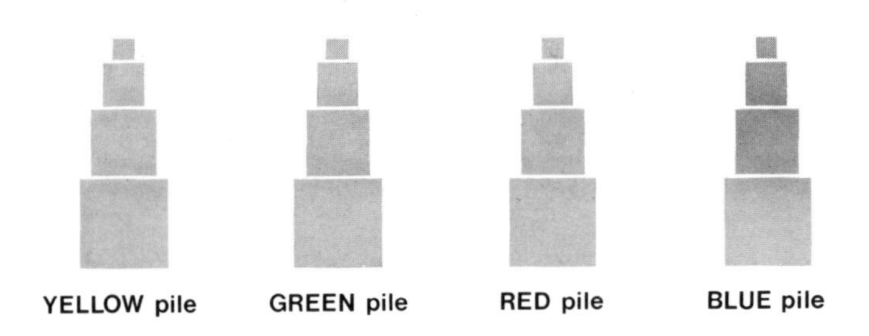

To a parental onlooker, though perhaps not to Peter, this will appear as an equally satisfactory arrangement.

Suppose now that Peter had sorted the cubes not according to colour but according to size, and finished up with:

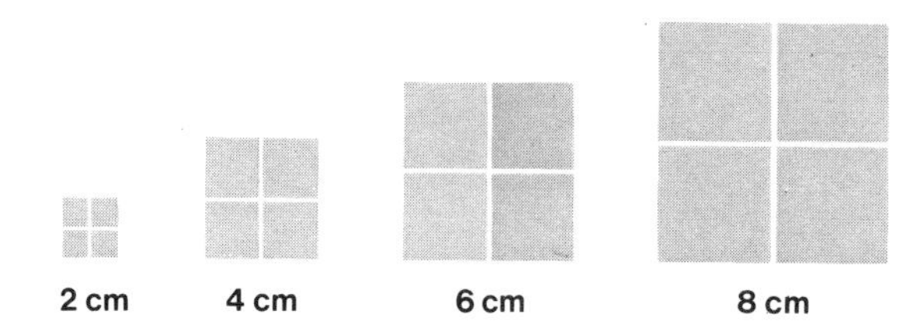

Suppose Paul changes the piles to:

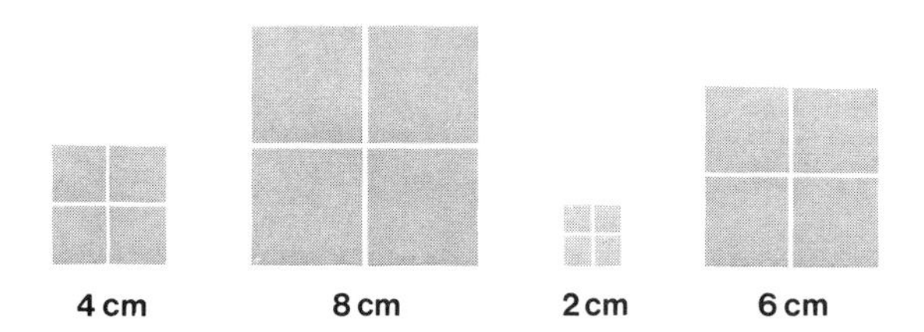

Then, if Peter assaults Paul, a parental observer may agree that there is some provocation because the final arrangement of the piles is now "wrong", and this is something that Peter can somehow understand even if his lack of counting ability and vocabulary make it difficult for him to put his objections precisely into words.

Of course, we can think up examples where a definite order is imposed upon colours (in a rainbow or traffic signal, for example), but, in general, there is a sense in which "sorting according to colour" is a less exacting process than "sorting according to size". When, therefore, we come to make mathematical models of these two types of process, we should expect to find a difference in the models. The difference in the processes is highlighted if we suppose that Peter's cubes are all of different colours and all of different sizes. Sorting the bricks according to colour is then a trivial process, but ordering them according to size is not.

Let us now look at a more sophisticated example: the case of the librarian classifying books according to subject matter and then arranging them on shelves in alphabetical order of authors within each classification. Here, there is both a *classification* and an *ordering*, one superimposed upon the other as it were. To be more specific and yet simple, suppose that our librarian simply classifies books into three sections, Fiction, Non-Fiction and Reference, and arranges the books within these sections in alphabetical order of authors. A possible arrangement of his bookcases would be:

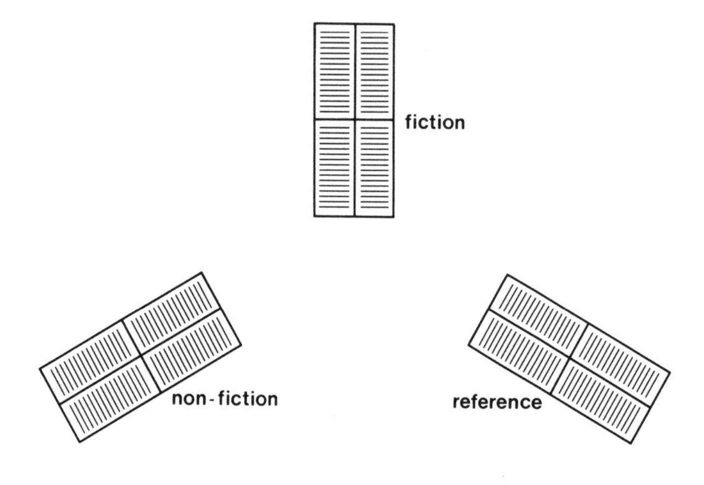

It clearly does not really matter if, in the course of redecoration of the library, the bookcases get moved around and finish up as:

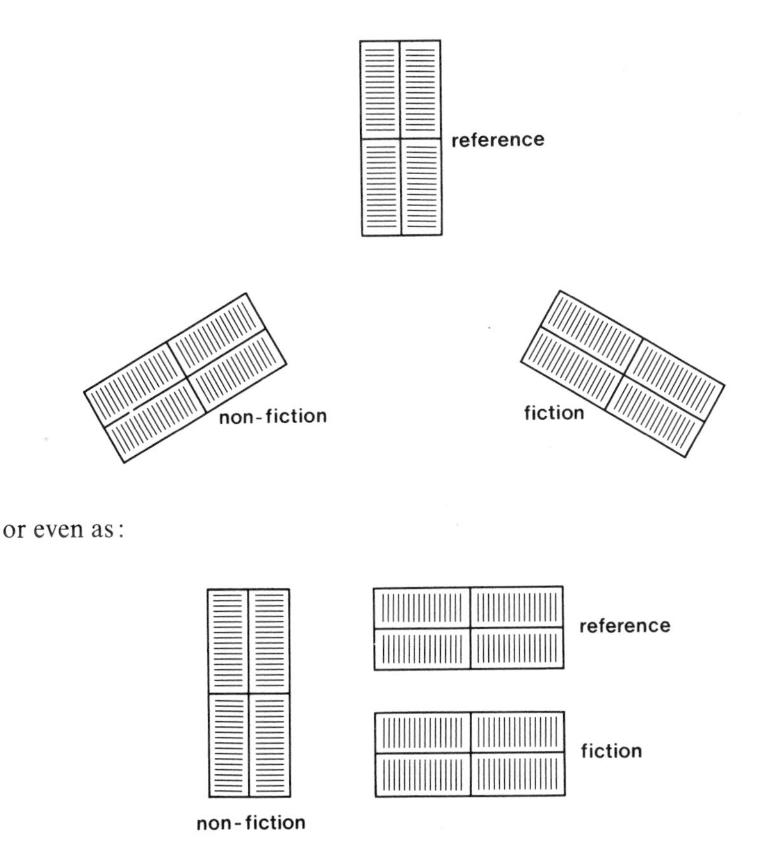

or even as:

But it will certainly matter a great deal if one of the bookcases is knocked over and the books are spilled on the floor, and then just put back anyhow, regardless of their original positions.

To put this in a slightly different way, we can say that if the librarian is faced with two books from the same section, say *Oliver Twist* by Charles Dickens and *Sons and Lovers* by D. H. Lawrence, then he will immediately know that the former should appear in the fiction cases "before" the latter. But supposing our librarian is faced with *Oliver Twist* and *A History of Europe* by H. A. L. Fisher, then the question of which should come first on the shelves is no longer meaningful, because they do not belong to the same section of the library.

This last case is in contrast to the situation of ordering cubes according to their size, where, given any two cubes, we can immediately say either that they are of the same size or that one is smaller than the other.

All these situations have their counterparts in their corresponding mathematical models. Sometimes it is necessary for mathematicians to *assume* that, given any two elements of a particular set, we can always determine which element comes first according to some prescribed ordering. In fact, this assumption is needed to justify a great deal of mathematics, because there are sets of numbers which have not been *proved* to be ordered.

It is this kind of problem that justifies the need for describing *sorting* and *ordering* in precise mathematical language. The basic concepts may appear simple, but to apply them in abstract mathematical circumstances may be very far from simple. When we are in the abstract world of mathematics, we must be in a position to analyse the situation with mathematical rigour if we are to be able to ask meaningful questions and arrive at meaningful answers.

There are two other good reasons why we should be interested in mathematical models of the processes of sorting and ordering. One reason is that we classify our knowledge in mathematics and need to use this knowledge in a particular order so often that mathematical models of the processes act as recurrent unifying threads throughout mathematics. (For example, we order geometric theorems.) The second reason is that once we have set up models of situations which arise over and over again, it is both possible and worth while to give such models a life of their own, analysing them and extending them so as to give general results which are applicable in many different situations.

The previous paragraph is very important. This unit on *relations* introduces you to some basic concepts in mathematics, which recur in many branches of the subject.

Our starting point is once again a set: a collection of well-defined objects. The child playing with his bricks may begin the story, but by the end of this text we shall have developed the plot so far that you will be hard put to it to remember its apparently humble beginnings. (The same thing happened in our discussion on *differentiation*. We started by looking for greater clarity in what we mean by the *velocity* of a car, and then the whole subject rapidly developed so that the velocity of a car became only a small dot upon a broad mathematical landscape.)

19.1 RELATIONS

19.1

19.1.0 Introduction

19.1.0

Introduction
* *

In section 19.1 we shall begin by looking back to some by now familiar concepts which we introduced in *Unit 1, Functions* and in *Unit 3, Operations and Morphisms.* In both these units an essential ingredient is a well-defined collection of objects, which we call a *set.* The adjective *well-defined* is inserted to emphasize the essential basic requirement that we must always specify a set in such a way that, given any object whatsoever, we can determine whether or not that object is a member of the set in question.

As the heading to this section (and the title of the unit) indicates, we shall be considering *relations.* In our context the word retains much of its usual meaning, but we have, as always, to give it precision. When we say that two people are related, this is very imprecise. Some people may stop at cousins when considering relations; others may include a brother-in-law's aunt. So we shall start by looking at one or two mathematical examples and then proceed to definitions.

Example 1

Example 1

If a set B is a subset of a set A, then we write

$$B \subseteq A,$$

and if it is not a subset, then we write

$$B \nsubseteq A.$$

We can think of the subset property as defining a *relation* between sets: that is, we say that B is *related* to A if $B \subseteq A$. Thus, for the set of positive integers, Z^+, and the set of integers, Z, we can say

$$Z^+ \subseteq Z,$$

that is, Z^+ is *related* to Z, but

$$Z \nsubseteq Z^+,$$

that is, Z is not *related* to Z^+.

Here we are dealing with a particular relation, which we may put into words as "is a subset of" (cf. "is a brother of"). However, such a phrase *on its own* is not sufficient to specify a relation, because we need also to state the set or sets whose elements are being compared. For example, starting with a specified person, Fred Jones, the relationship "is the father of" might give completely different answers if it were defined on the set of all people to those obtained if it were defined on the set of all males. ■

Example 2

Example 2

In *Unit 3* we defined an *operation*, and in particular, a *binary operation* on a set. You will recall that a binary operation combines any two elements of a given set so as to produce another element. For example, the binary operation of multiplication on Z gives us

$$7 \times 3 = 21.$$

We shall say that the integer 21 is *related* to the ordered pair of integers (7, 3) by a relation which we may put into words as "is the product of". Notice again that we must specify the sets from which we are comparing elements, in this case the set Z and the set $Z \times Z$ (the Cartesian product of Z with itself). ■

Example 3

Example 3

In *Unit 1* we defined a *mapping*, and in this case we could say that an element of the codomain is *related* to an element of which it is an image by a relation which we can put into words as "is an image of" (under the appropriate mapping). Again, for the relation to be properly defined we must know the sets whose elements are being compared; these could be the codomain and domain. ■

We have looked at these three examples because we want to highlight that a *relation* is something *very general indeed*, and because of this generality, relations play an important role in mathematics. In this section we begin by discussing what mathematicians mean by a *relation*, and then we go on to look at certain properties which some relations have and others do not, in order to be able to distinguish between different kinds of relation.

19.1.1

19.1.1 Definition of a Relation

Before formally defining a *relation* we shall consider one further example.

Example 1

Example 1

Consider the following two sets of names of men and of women respectively:

A = {Jim, Fred, Tom, Arthur},
B = {Mary, Anne, Sarah, Karen, Jane}.

We can compare elements from these sets by, for example, considering which (if any) of the men are married to one of the women. If we start with a member of set A and compare this member with the members of set B under the relationship which we may express in words as "is the husband of", we can make a list of statements, such as this:

Jim is not the husband of Mary,
Jim is the husband of Anne,
Jim is not the husband of Sarah,

...

We can then consider Fred's relationship with the women and obtain:

Fred is not the husband of Mary,
Fred is not the husband of Anne,

...

We shall perhaps find that Fred is not the husband of any of the women named in set B. We can continue the list until we have exhausted all the possibilities. We shall obtain *two sets of ordered pairs*; one of pairs for which the given relationship holds, and one of pairs for which the relationship does not hold.

We could, for example, obtain a set of married couples:

{(Jim, Anne), (Tom, Mary), (Arthur, Jane)},

and a set of unmarried pairs:

{(Jim, Mary), (Jim, Sarah), (Jim, Karen),
(Jim, Jane), (Fred, Mary), ..., (Arthur, Karen)}.

At a first glance, it may appear that in our example all we are doing is describing some particular kind of mapping. In fact a relation is *more general* than a mapping. ■

Exercise 1

Exercise 1
(2 minutes)

Why does the set of ordered pairs

{(Jim, Anne), (Tom, Mary), (Arthur, Jane)}

not define a mapping with domain the set A and codomain the set B? ■

Discussion
* *

We see that *a mapping can be expressed as a relation, but a relation is not necessarily a mapping.*

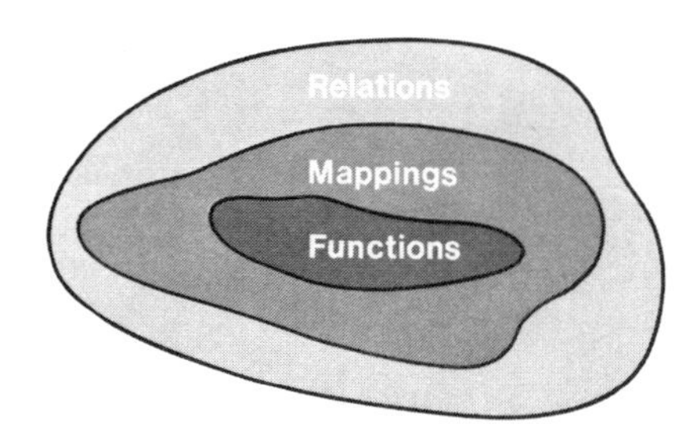

Returning to Example 1, we made the point that we ended up with two sets of ordered pairs. It is important to notice that the union of these two sets exhausts all the possible pairs, that is to say, it is equal to the Cartesian product, $A \times B$. You should also notice that their intersection set is the empty set. This is not just an accident due to the particular example we have chosen, as we shall see later.

Next we look at a more abstract example.

Example 2

Example 2

Consider the diagram:

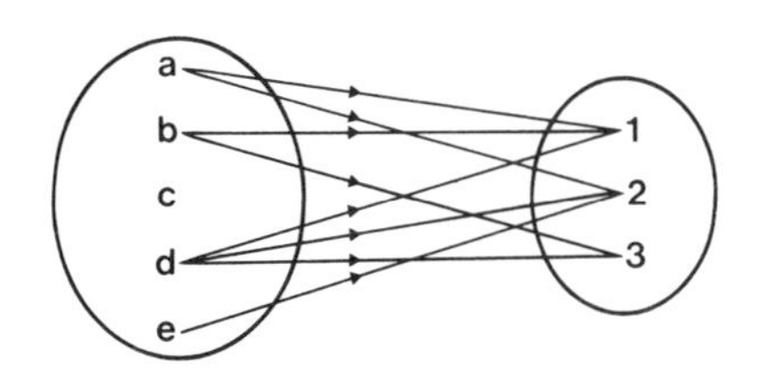

We can again compile two sets of ordered pairs by considering which letters are linked to which numbers. The relationship can be expressed in words as "is linked by a line to", but we shall adopt a more general notation and write

Notation 1
* * *

$a \rho 1$

to show that a is related to 1, and

$a \not\rho 3$

to show that a is not related to 3 (just as we write $x \in A$ if element x is a member of set A, and $x \notin A$ if x is not a member of A).

We use the Greek letter ρ (rho) instead of the more natural R, because we are already using R for the set of real numbers. We shall refer to *the relation* ρ.

We see from the diagram that

$a \rho 1$

$a \rho 2$

$a \not\rho 3$

$b \rho 1$, etc.

(*continued on page 8*)

Solution 1

Solution 1

The set of ordered pairs does not define a mapping because there is no "image" of Fred. The definition of a mapping requires that *each* element of the domain be mapped to an image in the codomain. ■

(*continued from page 7*)

Our two sets of ordered pairs are:

$$\{(a, 1), (a, 2), (b, 1), (b, 3), (d, 1), (d, 2), (d, 3), (e, 2)\},$$

the set for which ρ holds, and

$$\{(a, 3), (b, 2), (c, 1), (c, 2), (c, 3), (e, 1), (e, 3)\},$$

the set for which ρ does not hold.

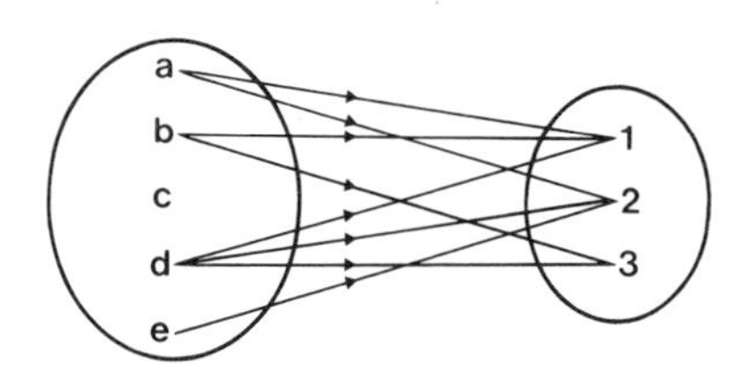

You should notice that the union of these two sets is equal to the Cartesian product

$$\{a, b, c, d, e\} \times \{1, 2, 3\},$$

and that the intersection of the two sets is the empty set. ■

The set of ordered pairs for which some given relationship holds is called the solution set of the relation concerned. (We have met the idea of a solution set before, in *Unit 6, Inequalities.*) Thus, in Example 1, the solution set is

Definition 1
* * *

$$\{(\text{Jim, Anne}), (\text{Tom, Mary}), (\text{Arthur, Jane})\},$$

and in Example 2 the solution set is

$$\{(a, 1), (a, 2), (b, 1), (b, 3), (d, 1), (d, 2), (d, 3), (e, 2)\}.$$

Discussion
* *

Notice that we have been careful to refer to "is the husband of" and "is linked by a line to" as *relationships* and not relations. This is because we want to emphasize that a relation is not defined unless we are given the set or sets concerned, just as a function is not defined just by the rule for obtaining the images, for we also need to know the domain and a co-domain.

If we consider the two examples above carefully, we see that we have what we might call a "chicken and egg" situation, a situation which often arises in mathematics.

We have two possible ways of approaching the definition of a *relation.*

(i) We may define a relation by specifying the set of ordered pairs for which the relationship concerned holds (i.e. the solution set), together with the set for which it does not hold.
(ii) Alternatively, we may define a relation by specifying two sets (which may be equal), together with a statement in words or symbols of the relationship by which we compare elements of one set with elements of the other.

Whichever method we choose, the other defines the *same* relation. There is a third, hybrid possibility. We can specify two sets, together with a subset of their Cartesian product such that this subset is the solution set of the relation.

Example 3

Example 3

A relation from set A to set B is defined by

$$A = B = \{2, 3, 4, 8\},$$

together with a relationship:

$a \rho b$ if and only if a is a multiple of b, i.e. there is an integer n, $n \neq 1$, such that $a = n \times b \quad (a \in A, b \in B)$. ■

Example 4

Example 4

A relation is defined by the solution set:

{(Shakespeare, *Hamlet*), (Shakespeare, *Othello*), (Shaw, *St. Joan*), (Shaw, *The Apple Cart*)},

together with the set:

{(Shakespeare, *St. Joan*), (Shakespeare, *The Apple Cart*), (Shaw, *Hamlet*), (Shaw, *Othello*), (Rattigan, *Hamlet*), (Rattigan, *Othello*), (Rattigan, *St. Joan*), (Rattigan, *The Apple Cart*)}. ■

Exercise 2

Exercise 2
(3 minutes)

(i) Define the relation of Example 3 in terms of sets of ordered pairs.
(ii) Define the relation of Example 4 in terms of sets and a relationship ρ in words. ■

Main Text
* *

Of course, it is not always practicable to list all the elements of the sets of a relation, nor to list all the ordered pairs of the solution set. But whichever definition of a relation we adopt, we must make clear both the relationship and also the set or sets to which it applies. We now give two possible equivalent definitions of a relation.

Definition 2a
* * *

Given two sets A and B (which may be equal), any subset* of $A \times B$ defines a relation from A to B. If (a, b) belongs to this subset, then a is related to b. If (a, b) is an element of $A \times B$ and does not belong to this subset, then a is not related to b.
(*The subset may be empty.)

Definition 2b
* * *

A relation is defined by two sets (which may be equal), together with a statement which is either true or false when it is used to link any member of one set with any member of the other set in a prescribed order.

(The words *prescribed order* are necessary, because the statement may be, for example, "a is taller than b"; it is clearly important that a is always drawn from one set and b from the other.)

We may ask ourselves which of the two definitions is to be preferred. The answer is that it depends on the particular way in which our information presents itself. It is useful to have the two alternative definitions at our disposal.

Example 5

Example 5

It is sometimes useful to represent the solution set of a relation diagrammatically. In *Unit 6, Inequalities* we looked at expressions of the type:

$$x^2 + y^2 \geqslant 1 \qquad (x \in R, y \in R).$$

This expression can be considered as defining a relation. In the sense of Definition 2a we can take $A = R$, $B = R$, and the subset of $R \times R$ as

$$\{(x, y) : x^2 + y^2 \geqslant 1, x \in R, y \in R\}.$$

(continued on page 10)

Solution 2 **Solution 2**

(i) Solution set:
$\{(4, 2), (8, 2), (8, 4)\}$.
Other set:
$\{(2, 2), (2, 3), (2, 4), (2, 8), (3, 2), (3, 3), (3, 4), (3, 8), (4, 3), (4, 4),$
$(4, 8), (8, 3), (8, 8)\}$.

Notice that it is not sufficient simply to specify the solution set, as that set gives no indication that 3 is one of the numbers being considered.

(ii) $A = \{$Shakespeare, Shaw, Rattigan$\}$
$B = \{$*Hamlet, Othello, St. Joan, The Apple Cart*$\}$
$a \rho b$ ($a \in A, b \in B$) if a is the author of b.

Notice again that had we not given the second set in Example 4 we would not have known that Rattigan was one of the authors belonging to set A.

■

(*continued from page 9*)

In the sense of Definition 2b the two sets are both R, and a possible statement would be

$$x^2 + y^2 \geqslant 1, \text{ where } x \in R \text{ and } y \in R.$$

We can represent the relation by the following diagram:

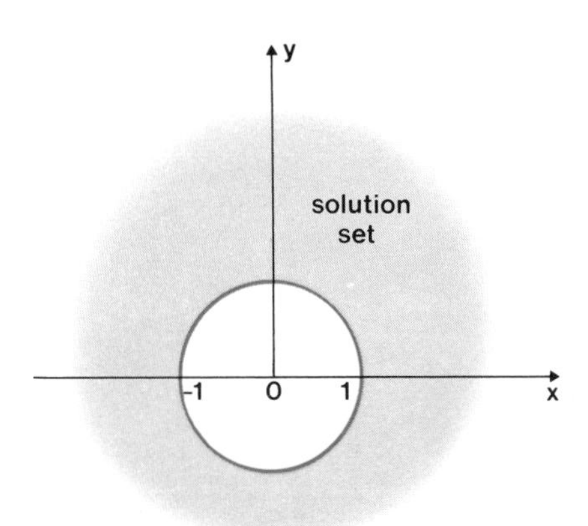

The solution set, a subset of $R \times R$, is represented by the set of all points on and outside the circle defined by $x^2 + y^2 = 1$. ■

You will probably have noticed that we have defined a relation so far solely in terms of the comparison of *two* elements. Strictly speaking, what we have been discussing are *binary relations*, and there are also such things as *ternary, . . . , n-ary, etc. relations* just as there are ternary, . . . , n-ary, etc. operations. By means of an *n-ary operation* (as we have seen in *Unit 3*) we obtain a result from an ordered n-tuple of elements of a set. Thus a closed n-ary operation on a set A is a function:

$$\underbrace{A \times A \times \cdots \times A}_{n \text{ terms}} \longrightarrow A.$$

By an *n-ary relation*, on the other hand, we accept or reject ordered n-tuples of elements which come from n sets (not necessarily all different).

Example 6

Example 6

Let A be the set of all candidates for some particular examinations; let B be the set {passed, failed, did not sit}; let C be the set of all subjects taken. We can now define a *ternary relation* where the corresponding relationship in words is:

"candidate a passed/failed/did not sit the examination in subject c".

If (say) Jones passed in Chemistry but failed in Mathematics, we should include in the solution set the ordered triples (Jones, passed, Chemistry) and (Jones, failed, Mathematics), but exclude the ordered triples (Jones, passed, Mathematics), (Jones, failed, Chemistry), (Jones, did not sit, Chemistry), (Jones, did not sit, Mathematics). ■

Main Text
* *

In the remainder of this text we shall confine our discussions to binary relations, and so we shall, in general, drop the adjective *binary*. We shall also concentrate on relations from a set A to a set B, where A and B are the same set. In these cases, instead of writing about relations from A to B, we shall in general discuss relations on A.

In the next section, we shall examine some particular types of relation and then describe them in abstract terms. In distinguishing between various relations we shall bear in mind that we are particularly interested in *sorting* and *ordering*.

19.1.2 Types of Relations

19.1.2

Discussion
* *

We began this text by considering the problem of sorting elements of sets into subsets. We are seeking to abstract the essentials of this process, and it is partly to this end that we have introduced the idea of a *relation*. We shall see that some relations do indeed sort elements in this way, whilst others do not; that is, different types of relations have different properties.

In order to single out certain properties which hold for some relations and not for others, and thus be able to classify relations according to such properties, we shall now look at a number of examples.

Example 1

Example 1

A relation on the set of integers, Z, expressed by

x is related to y if $x < y$ $\quad (x, y \in Z)$.

The solution set is a subset of $Z \times Z$, and corresponds to the following set of points:

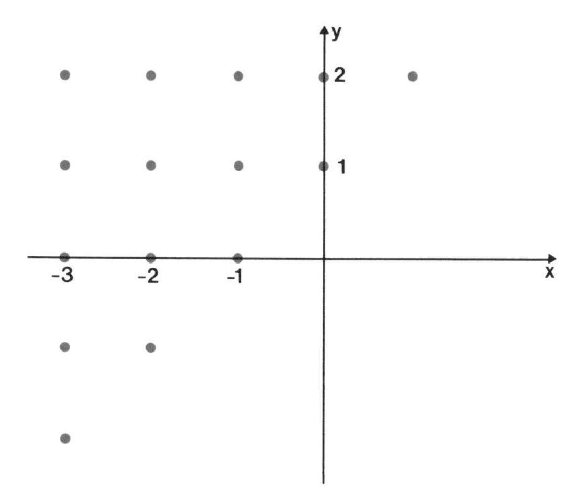

■

Example 2

Example 2

A relation on the set of integers, Z, expressed by

x is related to y if $x \leqslant y \quad (x, y \in Z)$.

In this case we get the following representation of the solution set:

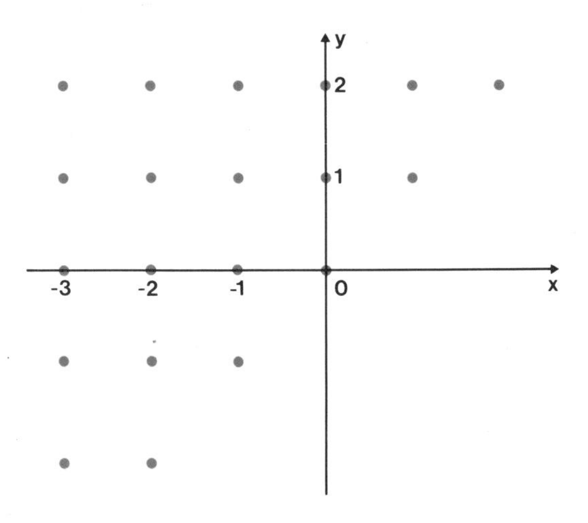

■

These two relations have already been discussed in *Unit 6*, but there is a particular difference between them which we pointed out there, and now want to highlight. To do this, we consider ordered pairs of integers (x, x).

Looking at Example 1, we see that pairs such as (x, x) would be rejected from the solution set of the relation. In the case of the relation of Example 2, however, all ordered pairs of the form (x, x) would be accepted.

Example 3

Example 3

Consider the diagram:

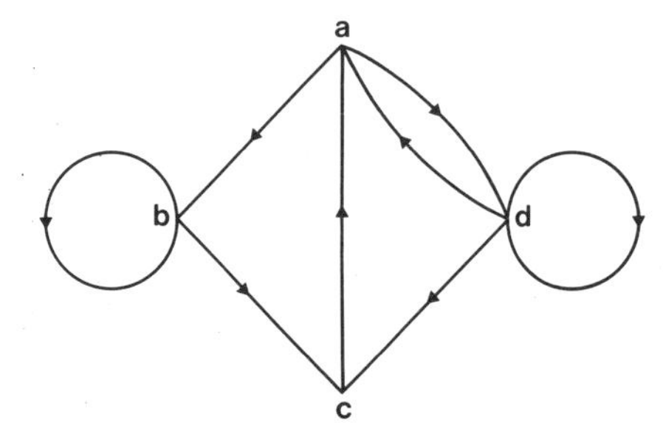

It consists of a number of points, labelled a, b, c and d, which are linked by paths, where the direction is indicated for each path. The paths are all, as it were, one-way streets. We can define a relation on the set of points $\{a, b, c, d\}$ by the sentence:

"x is related to y if there is a direct path from x to y $(x, y \in \{a, b, c, d\})$".

(By a *direct* path we mean a path which does not pass through any of the other labelled points.)

Let us again see what happens to ordered pairs of the form (x, x). There is no direct path from a to a, for example, as once we leave a we must go first to either b or d. There is, however, a direct path from b to b.

Thus the ordered pair (a, a) is rejected from the solution set of the relation, whereas the ordered pair (b, b) is accepted. ■

Comparing the three examples, we have first a case where *no* pair (x, x) is accepted, secondly a case where *all* pairs (x, x) are accepted, and thirdly a case where *some* pairs (x, x) are accepted and some rejected. We can write this as follows:

In Example 1, $x \rho x$ for *no* $x \in Z$.
In Example 2, $x \rho x$ for *all* $x \in Z$.
In Example 3, $x \rho x$ for *some*, but not all, $x \in (a, b, c, d\}$.

We now give a name to the property we have been considering.

Definition 1 * * *

A relation on a set A is reflexive if (x, x) belongs to the solution set of the relation for *every* $x \in A$.

Alternatively, a relation ρ on a set A is reflexive if

$$\forall_x \, x \rho x \; (x \in A)^*.$$

Thus, Example 2 is a reflexive relation, but Examples 1 and 3 are not. The term *reflexive* comes from the ordinary English word *reflex*, meaning *directed back.*

Exercise 1

Exercise 1 (3 minutes)

State whether or not the following relations are reflexive. For instance, in the first case, no person is taller than himself, so x is not related to x for any x in the set, and the relation is not reflexive.

(i) x is related to y if x is taller than y, on the set of Open University students.

(ii) x is related to y if x is the father of y, on the set of all people born in Great Britain.

(iii) x is related to y if x and y are the same height measured to the nearest inch, on the set of Open University students.

(iv) x is related to y if x and y are made by the same motor company, on the set of all motor cars registered in Great Britain in 1971.

(v) x is related to y if x and y are both manufactured by B.L.M.C., on the set of all motor cars registered in Great Britain in 1971.

(vi) x is related to y if x and y have the same derived function, on the set of all differentiable functions. ■

Many mathematical situations exhibit symmetry. To look for another distinguishing feature between relations we now ask:

If (x, y) is a member of the solution set, is (y, x) also a member?

Consider the following example of a relation.

Example 4

Example 4

We can specify a relation on the set of integers, Z, by:

$$x \text{ is related to } y \text{ if } |x - y| < 4 \qquad (x, y \in Z).$$

(Remember that $|x|$ is the numerical value of the number x, disregarding its sign.)

We see that the answer to the question in this case is YES, since $|x - y|$ is equal to $|y - x|$. The solution set of this relation is shown in red in the following diagram. Noticing that (y, x) is the "reflection" of (x, y) in the line with equation $y = x$, we can see that if (x, y) belongs to the set, so does (y, x).

* $\forall_x$ is the *universal quantifier* (read as "for *all x*"), defined in *Unit 17*.

(*continued on page 14*)

Solution 1

Solution 1

(ii) Not reflexive. No person is his own father.
(iii) Reflexive.
(iv) Reflexive.
(v) Not reflexive. A Ford is not related to itself.
(vi) Reflexive. ■

(*continued from page 13*)

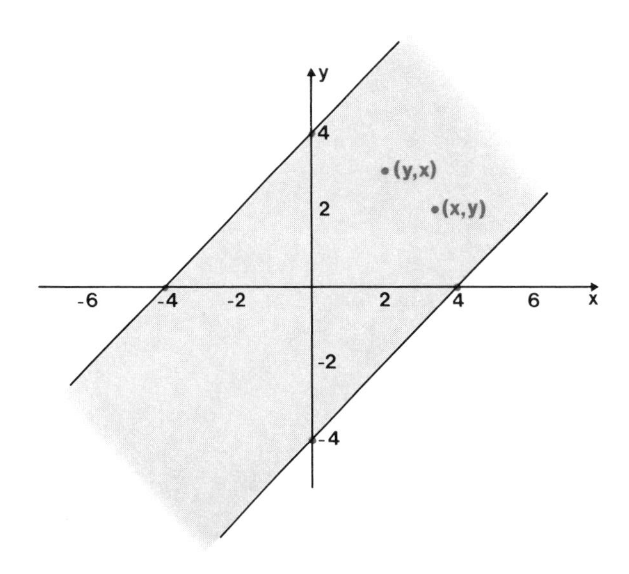

■

Looking back to the earlier examples, however, we find:

(i) in Example 1, where

$$x \rho y \text{ if } x < y \qquad (x, y \in Z),$$

that for no x such that $x \rho y$ is $y \rho x$;

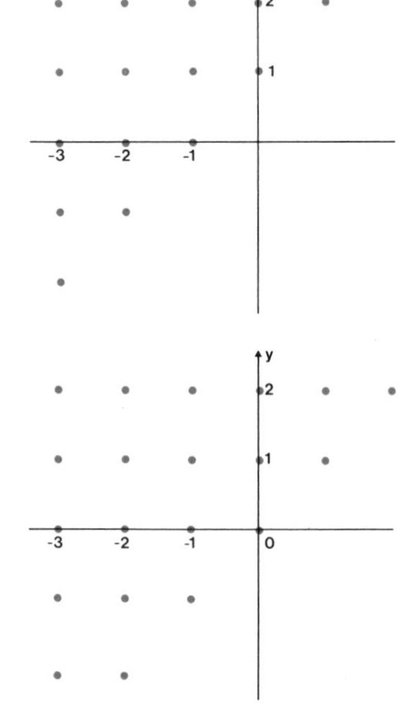

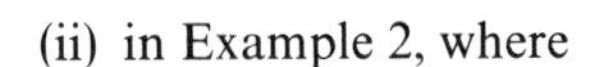

(ii) in Example 2, where

$$x \rho y \text{ if } x \leqslant y \qquad (x, y \in Z),$$

that if $x \rho y$, then $y \rho x$ only if $x = y$.

(iii) in Example 3, where

$$x \rho y \text{ if there is a direct path from } x \text{ to } y \quad (x, y \in \{a, b, c, d\}),$$

that $a \rho d$ and $d \rho a$, but otherwise at most one of (x, y) and (y, x) belongs to the solution set.

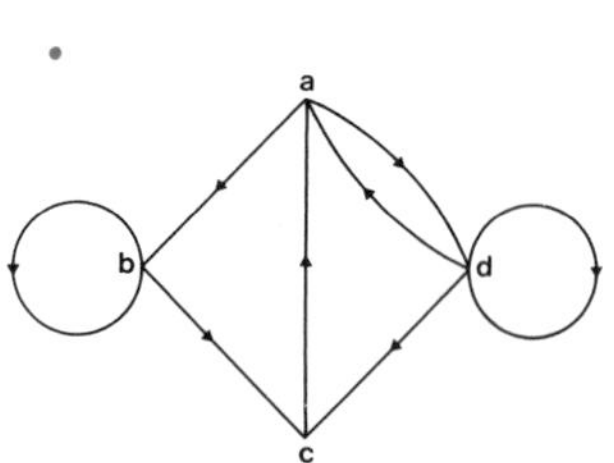

We again give a name to the property we have been considering.

Definition 2

A relation on a set A is symmetric if (y, x) belongs to the solution set *whenever* (x, y) belongs to the solution set $(x, y \in A)$.

Alternatively, a relation on a set A is symmetric if

$$\text{whenever } x \rho y, \text{ then } y \rho x, \qquad (x, y \in A).$$

Thus, of Examples 1–4 above, only Example 4 is a symmetric relation. The word *symmetric* refers to the property that the solution set is unchanged if the order of the elements in every pair is changed.

Exercise 2

Exercise 2
(3 minutes)

State whether or not the following relations are symmetric.

(i) Cases (i)–(iv) of Exercise 1 on page 13.

(ii) x is related to y if there is a 1 in row x and column y of the following table, on the set $\{a, b, c, d, e\}$. For example, looking along the top row (the "a" row) we have $a \rho a$, $a \rho b$, $a \rho e$, and $a \not\rho c$, $a \not\rho d$.

	a	b	c	d	e
a	1	1	0	0	1
b	1	0	1	1	0
c	0	1	0	0	1
d	0	1	0	0	0
e	1	0	1	0	1

(iii) As in (ii) but with the following table:

	a	b	c	d	e
a	1	1	1	1	1
b	0	1	0	0	0
c	1	1	1	1	1
d	0	0	0	1	0
e	1	1	1	1	1

(iv) x is related to y if the point with co-ordinates (x, y) lies on the circle with centre the origin and radius 1, on the set R.

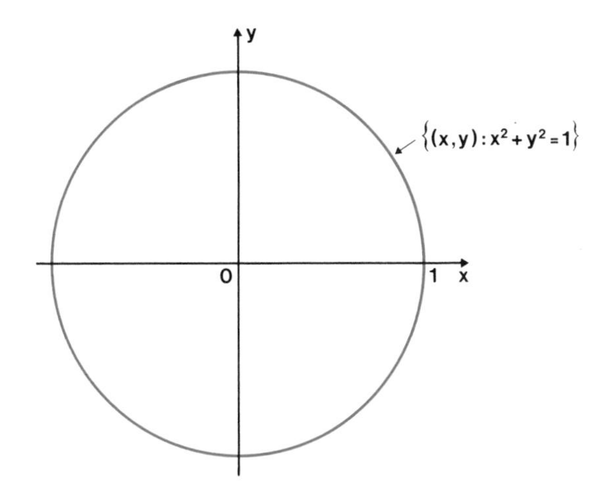

■

(*continued on page 17*)

Solution 2

Solution 2

(i) (i) No.
(ii) No.
(iii) Yes.
(iv) Yes.

(ii) Yes. The term *symmetric* is visually demonstrated in this case by the fact that the table is symmetric about the diagonal from the top left-hand corner to the bottom right-hand corner.

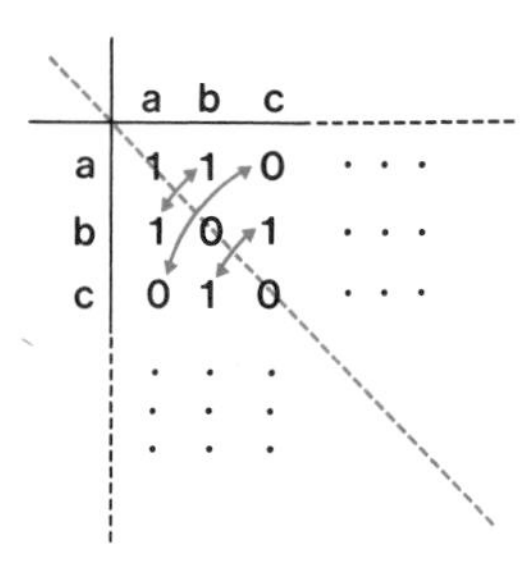

	a	b	c	...
a	1	1	0	...
b	1	0	1	...
c	0	1	0	...
...	...	...	...	

(iii) No. The table is not symmetric about the diagonal from top left to bottom right, and, for instance, $a \rho b$ but $b \not\rho a$.

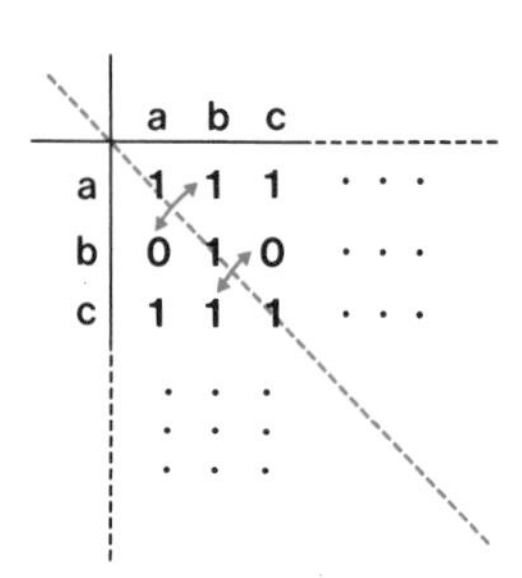

	a	b	c	...
a	1	1	1	...
b	0	1	0	...
c	1	1	1	...
...	...	...	...	

(iv) Yes. The circle is symmetric about the line with equation $y = x$. Algebraically, if $a \rho b$, then $a^2 + b^2 = 1$. This implies that $b^2 + a^2 = 1$, and so $b \rho a$.

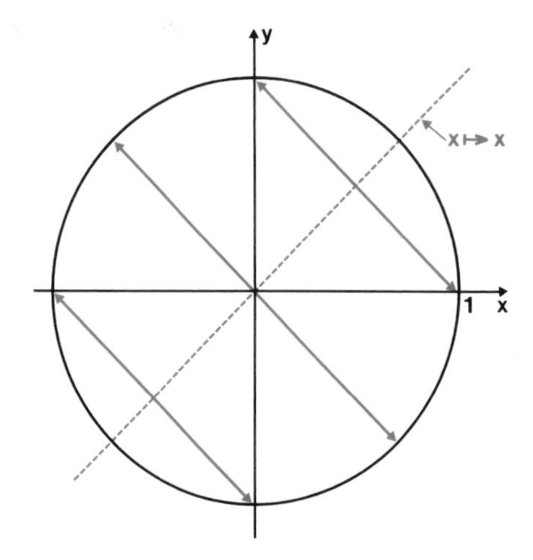

■

(*continued from page 15*)

Discussion
* *

We now have two properties of relations for which to look; the *reflexive* and *symmetric* properties. To consider the next property, we go back to Examples 1 and 2, namely the relations

Example 1

$$x \rho y \quad \text{if } x < y \qquad (x, y \in Z)$$

and

Example 2

$$x \rho y \quad \text{if } x \leqslant y \quad (x, y \in Z).$$

We have seen that neither of these relations is symmetric. We shall now define a property which expresses this in a more positive way.

Definition 3
* * *

A relation on a set A is anti-symmetric if *whenever* (x, y) and (y, x) both belong to the solution set, then x and y are the same element.

Alternatively, a relation on a set A is anti-symmetric if

$$x \rho y \text{ and } y \rho x \Rightarrow x = y \qquad (x, y \in A).$$

You may think at first sight that the relation of Example 1 is not anti-symmetric, but in the definition all we are saying is that pairs of the form (x, x) *may* belong to the solution set; they do not *have* to belong to it. Notice the word *whenever* in the definition; if (x, y) and (y, x) *never* both belong to the solution set, the condition is not violated.

It may seem to be implied by the terminology that it is impossible to have a relation which is both symmetric and anti-symmetric, but in fact it is possible. An example of such a relation is the relation on the set $\{0, 1, 2, 3\}$ with solution set

$$\{(0, 0), (1, 1), (2, 2)\}.$$

(You should check carefully that the conditions of both Definition 2 and Definition 3 are met.) This is why we use the term *anti-symmetric* rather than *not symmetric*, because it can be argued that in this special case we still have a rather trivial form of symmetry.

Both the relation of Example 1 and that of Example 2 satisfy the requirements of Definition 3, but the relation of Example 3, illustrated again in the following diagram, is neither symmetric nor anti-symmetric.

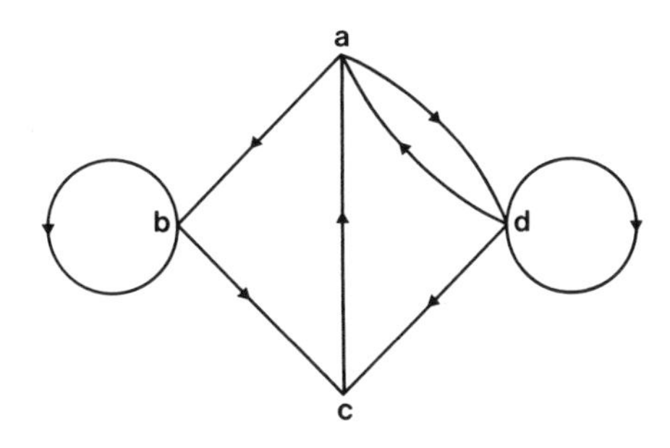

For instance, $a \rho b$ (because there is a direct path from a to b) but $b \not\rho a$, and so the relation is not symmetric. On the other hand, $a \rho d$ and $d \rho a$, and so the relation is not anti-symmetric either. The following examples both illustrate anti-symmetric relations.

Example 5

Example 5

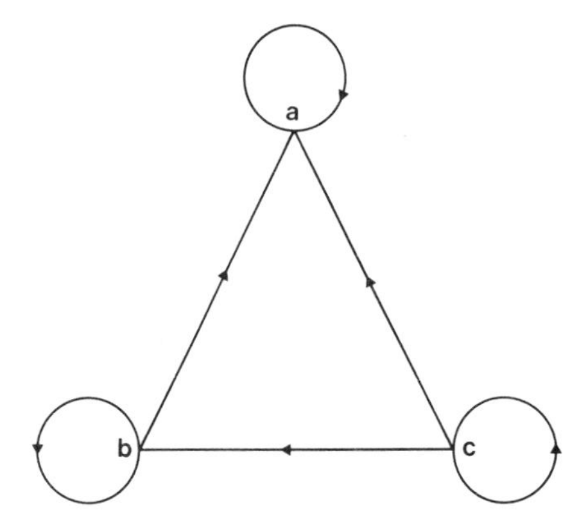

Set: the set of points $\{a, b, c\}$.

Relationship: there is a direct path from x to y $\quad (x, y \in \{a, b, c\})$. ■

Example 6

Example 6

Set: all the football teams in League Division 1 for the 1970–1971 season.

Relationship: team x is related to team y if x beats y at each meeting during the season.

(This relation is anti-symmetric in much the same way as the "$<$ relation" —there are no elements of the form (x, x) in the solution set.) ■

Main Text
* *

There is one further property that we wish to consider in this section. In *Unit 11, Logic I* we investigated the following tautology involving propositions:

$$[(\mathbf{a} \Rightarrow \mathbf{b}) \wedge (\mathbf{b} \Rightarrow \mathbf{c})] \Rightarrow (\mathbf{a} \Rightarrow \mathbf{c}).$$

In *Unit 6, Inequalities*, we saw that:

$$\text{if } a < b \text{ and } b < c, \text{ then } a < c \ (a, b, c \in R).$$

$\Rightarrow$ is a relation on a set of propositions, and $<$ is a relation on the set of real numbers, and we see here that they both have a similar property.

In general terms, the property is:

if a is related to b and b is related to c, then a is related to c.

Definition 4
* * *

A relation on a set A is transitive if (x, z) belongs to the solution set *whenever* (x, y) and (y, z) both belong to the solution set.

Alternatively, a relation on a set A is transitive if

whenever $x \, \rho \, y$ and $y \, \rho \, x$, then $x \, \rho \, z$.

The term *transitive* refers to the transition from x to z via the element y.

The proposition

$$[(\mathbf{a} \Rightarrow \mathbf{b}) \wedge (\mathbf{b} \Rightarrow \mathbf{c})] \Rightarrow (\mathbf{a} \Rightarrow \mathbf{c})$$

is a tautology precisely because the relation in the set of propositions defined by the connective of implication is transitive. Thus, if proposition **b** is implied by proposition **a**, and proposition **c** is in its turn implied by proposition **b**, it follows inevitably that proposition **c** is implied by **a**. For example, if "I jump off the bridge" implies "I fall in the river", and if "I fall in the river" implies "I get wet", then "I jump off the bridge" implies "I get wet".

The corresponding situation in set algebra is illustrated by the following Venn diagram:

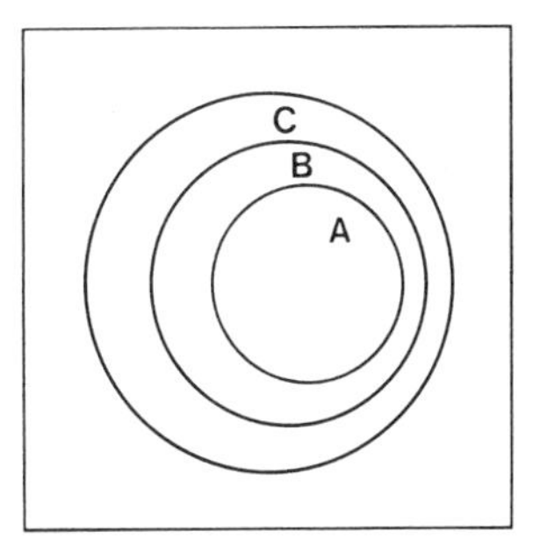

Since $A \subseteq B$ and $B \subseteq C$, it follows that $A \subseteq C$.

We see that the relation defined by the relationship

x is related to y if x is a subset of y,

on the set of sets $\{A, B, C\}$ is transitive.

On the other hand, the relation defined by the relationship

x is related to y if $x \cap y$ is not empty,

on the set of sets $\{A, B, C\}$, illustrated in the following diagram, is not transitive.

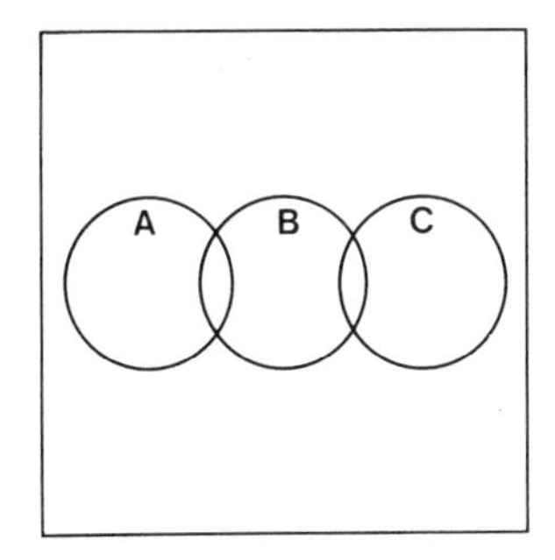

In this case we have $A \rho B$ and $B \rho C$, but $A \not\rho C$.

Another example of a transitive relation is the relation

"is parallel to"

on the set of all lines in a plane. For if a line L_1 is related to a line L_2, and if L_2 is related to a line L_3, then L_1 is related to L_3.

Exercise 3

Exercise 3
(3 minutes)

State whether or not the relations in the examples in Exercise 1 on page 13 are transitive. ■

Main Text
* *

We are now in a position to classify relations according to the reflexive, symmetric, anti-symmetric and transitive properties. We may seem to you to have journeyed some way from the concepts of *sorting* and *ordering* with which we began this text. In fact, we are now at the point where we can return to these concepts, because we can pick out two particular kinds of relation which will give us mathematical models of *sorting* and *ordering*.

(*continued on page 20*)

Solution 3

Solution 3

(i) Yes. If x is taller than y and y is taller than z, then x is taller than z.
(ii) No.
(iii) Yes.
(iv) Yes.
(v) Yes.
(vi) Yes.

At the end of this exercise you should review the reflexive, symmetric, anti-symmetric and transitive properties which relations may possess. You need to have a clear understanding of them for the subsequent discussion. ■

(*continued from page 19*)

Definition 5 * * *

A relation on a set A which is reflexive, symmetric and transitive is an equivalence relation on A.

Definition 6 * * *

A relation on a set A which is reflexive,* anti-symmetric and transitive is an order relation on A.

As an example of an equivalence relation, consider the set of all Open University students, and the relationship

x lives in the same Open University Region as y.

This is one of the relations which the administrative section of the University uses to sort the set of students. In section 19.2 we look closely at *equivalence relations* and see how they give us a mathematical model of the *sorting process.*

None of the earlier Examples 1–6 is an equivalence relation (you might like to see which of the required properties are lacking). The relations in (iii), (iv), and (vi) of Exercise 1 on page 13 are equivalence relations.

As far as *order relations* are concerned, the inequality relations are perhaps the most familiar examples — they enable us to arrange a set of numbers in "order of magnitude". Of our earlier examples, 1,* 2 and 5 are *order relations*; in each example we have a chain of elements, each element related only to all later ones. This is the familiar use of the word *order*. However, you will discover in section 19.3, where we examine order relations in more detail, that an order relation is much more general than you might think from these examples.

After the next two exercises we shall consider first *sorting* (in 19.2) and then *ordering* (in 19.3).

* It is usual in the mathematical literature to include the reflexive property in the definition of an order relation: notice that this means that $<$ on the set of integers is not strictly an order relation, although intuitively it orders the set. But if ρ is an order relation, then we can always define an associated "order" relation ρ_1, by

$$x \,\rho_1\, y \quad \text{if } x \,\rho\, y \quad \text{and} \quad x \neq y.$$

Then, for example, if ρ is $\leqslant$, ρ_1 is $<$, so we are not losing anything by the inclusion of the reflexive property in our definition.

Exercise 4

Exercise 4
(5 minutes)

Show that the following relations are equivalence relations. Suggest how the relations can be used to sort the sets into non-overlapping subsets.

(i) The relationship: $x \rho y$ if x and y have the same remainder on division by 3, on the set

$$\{1, 2, 3, 4, 5, 6, 7, 8\}.$$

(ii) The relationship: x is related to y if there is a direct path from x to y, on the set $\{a, b, c, d, e\}$, as shown in the following diagram:

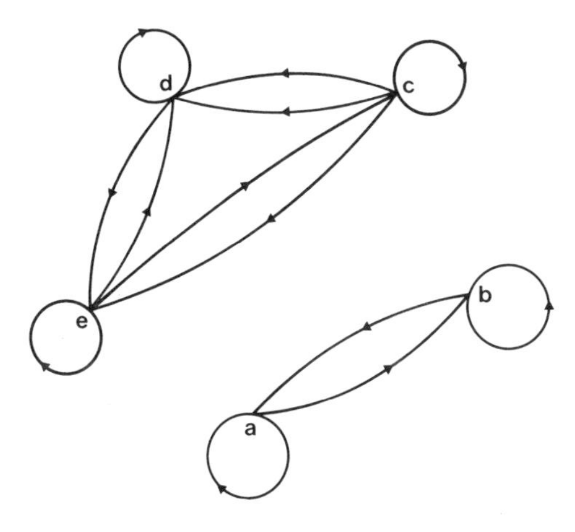

■

Exercise 5

Exercise 5
(5 minutes)

Classify the following relations as equivalence relations, order relations, or neither.

(i) x is related to y if there is a direct path from x to y on the set $\{1, 2, 3, 4, 5\}$, as shown in the following diagram:

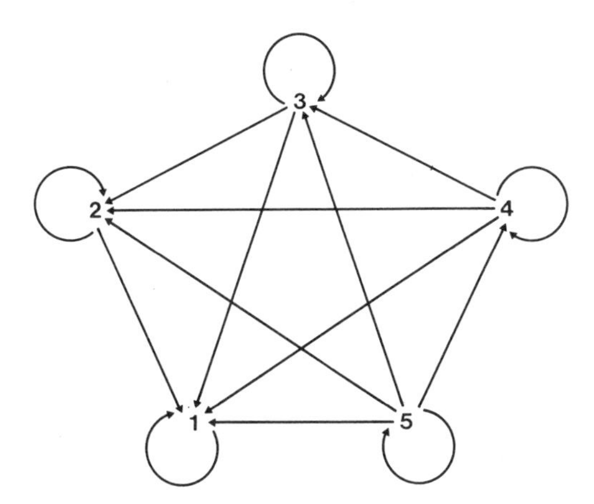

(ii) x is related to y if $\sin x = \sin y$ on the set of real numbers.

(iii) x is related to y if there is a 1 in row x and column y of the following table, on the set $\{a, b, c, d\}$.

	a	b	c	d
a	1	0	1	0
b	1	1	1	1
c	0	0	1	0
d	1	0	1	1

(iv) x is related to y if x is registered for a course for which y is also registered, on the set of all Open University students. ■

Solution 4

Solution 4

(i) $x \rho x$ for all x in the set (reflexive property).

If x leaves the same remainder as y on division by 3, then y leaves the same remainder as x. That is, if $x \rho y$, then $y \rho x$ (symmetric property).

If x leaves the same remainder as y, and y leaves the same remainder as z, then x leaves the same remainder as z. That is, if $x \rho y$ and $y \rho z$, then $x \rho z$ (transitive property).

The relation is reflexive, symmetric and transitive, and it is therefore an equivalence relation.

(ii) Each of the three properties can be checked as in (i).

We can sort the sets into non-overlapping subsets by collecting into subsets all the elements which are related to each other. For example, in (i) there will be one subset consisting of numbers which leave remainder 0, one of numbers which leave remainder 1, and one of numbers which leave remainder 2.

In (i) the set is sorted into three non-overlapping subsets of related elements, $\{3, 6\}$, $\{1, 4, 7\}$, $\{2, 5, 8\}$, and in (ii) the set is sorted into two subsets, $\{a, b\}$, $\{c, d, e\}$, of related elements. ■

Solution 5

Solution 5

(i) An order relation. The relation is, in fact, $\geqslant$. We can arrange the elements in order:

$$5, 4, 3, 2, 1,$$

so that each element is related to all the later elements.

(ii) An equivalence relation. It sorts the real numbers into subsets, every number in a particular subset having the same sine. For example, the subset $\{0, \pi, -\pi, 2\pi, -2\pi, \ldots\}$ consists of all the elements x such that $\sin x = 0$.

(iii) An order relation. The elements can be arranged in order: b, d, a, c, such that each element is related to all the later elements. Each element is related to itself, so the reflexive property is satisfied. Nowhere do we get $x \rho y$ and $y \rho x$ where $x \neq y$, so the anti-symmetric property is satisfied. By checking all the possible cases, we can see that the transitive property is also satisfied. For instance, b is related to d, d is related to a, and b is related to a.

(iv) Neither. For instance, x may be studying Social Science and Maths., y Maths. and Science, and z Science and Humanities, and so the relation is not transitive. (We are, of course, assuming that students registered for these combinations really exist.) ■

19.2 EQUIVALENCE RELATIONS

19.2

19.2.0 Introduction

19.2.0

Introduction
* *

In the Introduction to this text, we considered the case of the librarian classifying books according to subject matter. In practice, there are a number of complications in such a sorting process; for example, in what way should *A History of Chinese Mathematics* be classified? This is usually taken care of by some system of cross-referencing so that such a book appears in the library's index cards under History, under China, and under Mathematics. However, a single copy of the book cannot sit on the shelves of the library in three different places at the same time, so a decision has to be taken as to its exact location in the bookcases, whatever cross-referencing may be done. We shall neglect such complications and start by looking at the straightforward sorting process whereby books (of which we assume there are no duplicates) in a library are allocated to a main classification which will determine their location in the library.

First, it is clearly necessary that every book in the library shall be sorted into one of the classes; we cannot have books lying about on the librarian's desk simply because there is no classification which fits them. Equally well, we do not want to have odd books lying about of which the librarian has no knowledge. Put into mathematical terms, we need to *specify our set*, in this case the set of all books belonging to the library, and our classification system must be such that *each and every element of the set can be classified.*

But more than this is needed. We have already indicated that we have simplified our library situation so that each book is given a main classification which will determine its location in the library, and we have also pointed out that any given book can only occupy one place at a time. Put into mathematical terms, this means that *every element of our set must be uniquely classified.* In section 19.2.1 we look at the way in which a set must be broken down into subsets in order to meet the requirements of the sorting process.

19.2.1 Partitioning a Set

19.2.1

Main Text
* *

We begin by formally defining the word *partition.*

Definition 1
* * *

A partition of a set A is a separation of the elements of A into subsets such that each element is *in one and only one subset.*

Example 1

Example 1

Consider the set $\{a, b, c\}$. Possible partitions of the set are:

(i) $\{a, b, c\}$
(ii) $\{a\}, \{b, c\}$
(iii) $\{b\}, \{a, c\}$
(iv) $\{c\}, \{a, b\}$
(v) $\{a\}, \{b\}, \{c\}$

In each case, the union of the subsets gives us the original set, and the intersection of the (distinct) subsets is empty. For example, in the case of partition (iv), we have

$$\{c\} \cup \{a, b\} = \{a, b, c\}$$

and

$$\{c\} \cap \{a, b\} = \varnothing.$$

■

Main Text
* *

We now want to show how this idea of *partitioning* is directly linked with that of an *equivalence relation.*

Given a *partition* of a set A, we can define a relation on A by:

> $x \rho y$ if and only if x and y belong to the same subset of the given partition of A.

However we partition any given set A, this relation will necessarily be an equivalence relation, as can easily be checked. The transitive property is worth noting: if $x \rho y$ and $y \rho z$, i.e., if x belongs to the same subset as y and y belongs to the same subset as z, then we can conclude that $x \rho z$, i.e. that x belongs to the same subset as z, but only because we have insisted that the subsets of a partition *do not overlap.* If the subsets were allowed to overlap, then we could have as part of our partition

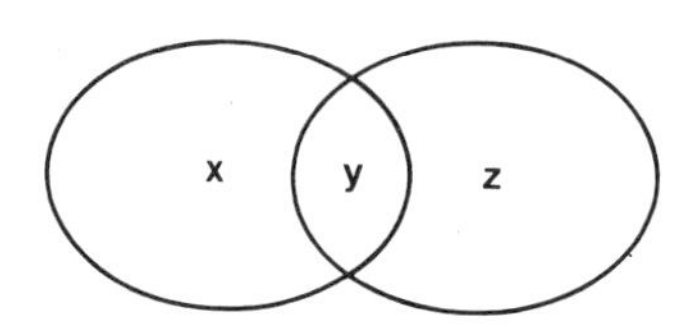

in which $x \rho y$ and $y \rho z$ but $x \not\rho z$.

Let us now look at the problem the other way round and start, not with a partition of a set A, but with an *equivalence relation* ρ on A.

We can now define subsets of A by:

x and y belong to the same subset of A if $x \rho y$.

This statement applies to all the elements of A in turn (because ρ is reflexive and so, at least, $x \rho x$), so each element must go into at least one subset of A. To show that we have a partition of A, we must show that no element of A belongs to more than one subset. Let us assume as a hypothesis the contradiction of the conjecture, that an element y belongs to two different subsets B and C, say (see *proof by contradiction*, section 17.2.4, *Unit 17, Logic II*). Suppose b is any element of B and c is any element of C.

We have $y \rho b$ and $y \rho c$.

But ρ is an equivalence relation, and is therefore reflexive, symmetric and transitive. By the symmetric property we have:

$$b \rho y$$

and from $b \rho y$ and $y \rho c$ we have, by the transitive property,

$$b \rho c,$$

and so b and c are related and therefore $b \in C$.

By the symmetric property,

$$c \rho b,$$

so $c \in B$.

We have shown that any element of B belongs to C and any element of C belongs to B; it follows that $B = C$, which contradicts the hypothesis that B and C are different.

So any equivalence relation on a set A partitions the set.

Definition 2
* * *

The subsets of the partition of a set A obtained from a given equivalence relation on A are called equivalence classes.

By allocating all the elements of a set uniquely to equivalence classes, we carry out in mathematical terms the process which in more general situations we have referred to as *sorting*. Thus, the mathematical model of the sorting process is the partitioning of a set by an equivalence relation into equivalence classes.

In practice, we frequently do not list all the elements of each equivalence class, and so we choose one particular element from each class as its representative. Any element belonging to a particular equivalence class may be chosen as its representative for the purpose of naming the class; for other purposes there may be external conditions which determine just how we make the choice.

Example 2

Example 2

Consider the equivalence relation on the set of real numbers expressed in decimal form:

x is related to y if $x = y$ when both are rounded* to four significant figures.

A typical equivalence class would be

$$\{x : x \in R \quad \text{and} \quad 3.1415 \leqslant x \leqslant 3.1425\}.$$

Here, it is natural to select the number 3.142 as a representative of the whole class, since all the numbers in the class are rounded to this number. Knowing the equivalence relation, we can decide whether or not any number belongs to the class just by knowing this (or any other) representative. ■

As we study more mathematics we shall find that the concept of an equivalence class and that of an equivalence relation recur frequently. As usual, having found a mathematical model of a fairly common (and important) procedure, we find that the model gets a life of its own and develops far beyond its origins. We shall see a little of this in this unit. It is also interesting to note that, using the concepts of equivalence relations and equivalence classes, we can examine parts of our mathematical experience and deepen our understanding of it.

Exercise 1

Exercise 1
(2 minutes)

Give the partition of $\{3, 4, 5, 6\}$ determined by the equivalence relation $x \rho y$ if x and y have the same remainder on division by n, where

(i) $n = 2$
(ii) $n = 3$
(iii) $n = 4$. ■

* See the convention given on page 7 of *Unit 2*.

Solution 1

Solution 1

(i) $\{4, 6\}, \{3, 5\}$
(ii) $\{3, 6\}, \{4\}, \{5\}$
(iii) $\{4\}, \{5\}, \{6\}, \{3\}$ ■

19.2.2 The Natural Mapping

19.2.2

Let us go back again for a moment to our librarian and his classification of books into three classes: Fiction, Non-fiction and Reference. When he decides to allocate a book to one of the three classes, he may well at the same time write an F, an N or an R on the spine of the book. He is thus, in effect, establishing a *mapping* from the set of all the books in his library to the set of classes denoted by F, N and R. This mapping is a *function* because, as we have seen earlier, in order to determine the part of the library in which to locate any individual book, no book is allocated to more than one class; that is, the image of each book is a unique equivalence class. For example, we might have:

Moby Dick $\longmapsto F$
Treasure Island $\longmapsto F$
Set Theory $\longmapsto N$
Cruden's Concordance $\longmapsto R$
Biographical Dictionary $\longmapsto R$
etc.

This is an example of a general situation of which the mathematical model is the mapping of the elements of a set A to the set of equivalence classes obtained from an equivalence relation on A. We call the set of equivalence classes the quotient set, and we denote it by A/ρ (which we read as "A by ρ"), where ρ represents the equivalence relation by means of which the classes are determined.

Definition 1
* * *

The notation A/ρ may be a little worrying; it suggests a connection with division, as indeed does the word *quotient*. A/ρ stands for a set of subsets of A, so an element of A/ρ will be a subset of A, namely one of the equivalence classes defined by ρ. However, we know that an equivalence relation partitions a set, and so we can think of ρ as "dividing up" the set into subsets. The process of assigning an element to its equivalence class specifies a mapping from the set A to the set A/ρ. Because we have a partition of A, each element of A has an image and the sets in A/ρ are non-overlapping, so the mapping is in fact a function.

The natural mapping,*

Definition 2
* * *

$$n: A \longmapsto A/\rho,$$

is the function which maps each element of a set A to its corresponding equivalence class under an equivalence relation ρ on A.

However, once again we have a "chicken and egg" situation, since the whole problem may arise the other way round. We may start with a function:

$$f: A \longmapsto B$$

in which case there is a natural equivalence relation on the domain A of f defined by

Definition 3
* * *

$$x \,\rho\, y \text{ if and only if } f(x) = f(y) \qquad (x, y \in A).$$

* We call it the *natural mapping*, rather than the *natural function*, to fit in with standard terminology: many authors define a *mapping* to be the same as a *function*.

Example 1

Example 1

Consider the function

$$f{:}x \longmapsto x^2 \qquad (x \in R).$$

Here, each equivalence class, except that containing only the number zero, has exactly two elements, a pair of numbers of equal magnitude but of opposite sign. ■

Example 2

Example 2

Consider the function

$$f{:}x \longmapsto \sin x \qquad (x \in R).$$

Here, each equivalence class contains an infinite number of elements, for example, the class of elements which map to zero under f is $\{0, \pi, -\pi, 2\pi, -2\pi, \ldots\}$. ■

19.2.3 Combination of Equivalence Classes

19.2.3

Main Text
* *

So far, we have discussed the partitioning of a set into equivalence classes, and the fact that such a partitioning leads to a particular mapping—the natural mapping. We shall now see that when a binary operation is defined on the set, it can sometimes be used to define an operation on the set of equivalence classes, a way of combining the classes themselves. The way we tackle the problem is to use the natural mapping, for we have already investigated the situations which arise when operations are defined on the domain of a mapping. We look to see if we get a *morphism.*

Example 1

Example 1

First consider the set of integers, Z. This set may be partitioned into those integers which are *odd* and those which are *even* (0 is considered to be even). If we call these classes O and E respectively, then the natural mapping, n, is a many-one mapping from Z to $\{O, E\}$.

For example

$$n{:}1 \longmapsto O$$

$$n{:}156 \longmapsto E$$

Now we introduce an operation — let us take multiplication as an example. It is easy to check that n is a morphism for multiplication on the set of integers and an operation □ on the set $\{O, E\}$ defined by the following table:

□	O	E
O	O	E
E	E	E

This table simply expresses the fact that an odd number multiplied by an odd number is odd, and so on. It represents a way of *combining the equivalence classes.* So we have used the operation $\times$ on Z to define a way of combining the classes themselves, and we shall be able to do this for any partitioning of a set S with an operation $\circ$, whenever the natural mapping is a morphism for $\circ$. ■

Exercise 1
(3 minutes)

Exercise 1

For the operation $+$ we can also construct a table which defines an operation on $\{O, E\}$ which makes the natural mapping a morphism. Construct this table. ■

Solution 1

Solution 1

$\square$	O	E
O	E	O
E	O	E

■

Main Text ★★

In *Unit 3, Operations and Morphisms* we said that a mapping is a morphism for an operation $\circ$ whenever the *mapping and* $\circ$ *are compatible.* A mapping f, with domain A, and an operation $\circ$ are compatible if, whenever

$$f(x_1) = f(x_2)$$

and

$$f(y_1) = f(y_2),$$

then

$$f(x_1 \circ y_1) = f(x_2 \circ y_2) \qquad (x_1, x_2, y_1, y_2 \in A).$$

The advantage of introducing this property is that it enables us to predict whether or not f is a morphism without having to find an operation in the codomain.

In the context of equivalence relations, it enables us to predict whether or not we can use an operation to define a combination of equivalence classes.

Of course, the natural mapping depends entirely on the equivalence relation, and so we extend the definition of the word *compatibility* and say that, when the natural mapping is a morphism for the operation $\circ$, the equivalence relation and the operation $\circ$ are compatible. The compatibility conditions can be expressed directly in terms of equivalence relations as follows.

Definition 1 ★★

Whenever

$$x_1 \rho x_2$$

and

$$y_1 \rho y_2,$$

then

$$(x_1 \circ y_1) \rho (x_2 \circ y_2) \qquad (x_1, x_2, y_1, y_2 \in A).$$

This simply means that, if we combine an element x from a class X with an element y from a class Y, then the result belongs to the *same* class, Z say, (the class containing $x \circ y$) for *all* choices of $x \in X$ and $y \in Y$. (For example, if $x_1, x_2 \in X$ and $y_1, y_2 \in Y$, then $x_1 \circ y_1$ and $x_2 \circ y_2$ both belong to Z.)

In the "odds and evens" example the classes are

$$\{\ldots, -6, -4, -2, 0, 2, 4, 6, \ldots\} = E$$

and

$$\{\ldots, -5, -3, -1, 1, 3, 5, \ldots\} = O$$

If we pick *any* number from E and multiply it by *any* number from O, we get a number in E, and so the compatibility conditions are satisfied and we are able to say that, corresponding to multiplication, O combines with E to give E.

The combination of equivalence classes which we obtain when the natural mapping is a morphism is defined as follows, where we denote the equivalence class containing x by $[x]$, and the operation in the domain is $\circ$:

Notation 1 ★★★

$$[x] \square [y] = [x \circ y].$$

Definition 2 ★★★

Since the operation $\square$ is in a sense an "extension" of $\circ$, we usually use $\circ$ to stand also for combination of classes, as in the following commutative diagram:

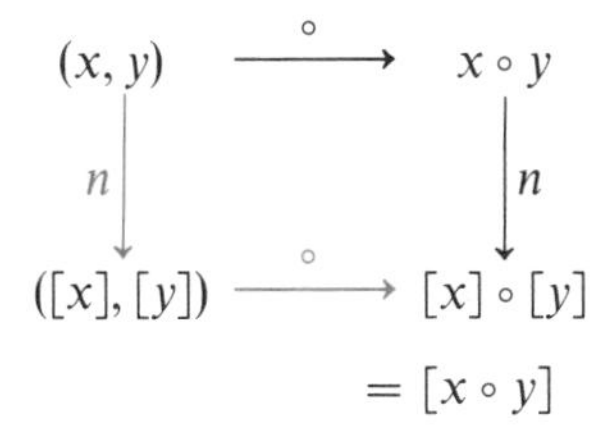

Example 2

Example 2

Consider the set of real numbers R in decimal form, and the equivalence relation on R expressed by:

> x is related to y if $x = y$ when both are rounded to four significant figures.

(We saw this example before on page 25.)

For the purpose of illustration we choose the two equivalence classes

$$S_1 = \{3.124, 3.1243, 3.1238, \ldots\},$$

$$S_2 = \{1.1604, 1.1597, \ldots\}.$$

Consider the binary operation of addition. We have, for example,

$$3.124 + 1.1604 = 4.2844$$

which is rounded to 4.284. We also have

$$3.124 + 1.1597 = 4.2837$$

which is also rounded to 4.284. So far, it looks as though the "sum" of the classes S_1 and S_2 is the class containing 4.284. But is it the case that *whatever pair of numbers we choose, one from each class*, their sum when rounded will always be 4.284? If we choose 3.1243 from S_1 and 1.1604 from S_2 we get

$$3.1243 + 1.1604 = 4.2847,$$

which does *not* belong to the class containing 4.284. The result depends on the representatives we choose for the classes, and so *we cannot extend the operation + to the set of equivalence classes.* The equivalence relation and the operation + are not compatible. ■

Exercise 2

Exercise 2
(5 minutes)

(i) Consider the equivalence relation on the set of real numbers:

> x is related to y if $x^2 = y^2$.

Is this relation compatible with
(a) addition?
(b) multiplication?

(ii) Consider the equivalence relation on the set of all functions with domain and codomain R:

> f is related to g if $f(0) = g(0)$.

Is this relation compatible with
(a) addition of functions?
(b) multiplication of functions?

(iii) Consider the equivalence relation on the set of all polynomial functions:

> f is related to g if f and g have the same derived function.

Is this relation compatible with
(a) addition of functions?
(b) multiplication of functions? ■

Solution 2

Solution 2

(i) (a) No.
(b) Yes.
You can get the answer to (a) by trying a few elements. The equivalence classes are $\{0\}, \{-1, 1\}, \{-2, 2\}$, etc. $-1 + 2$ does not belong to the same class as $1 + 2$. For (b), we can proceed algebraically as follows:

If x_1 is related to $x_2, x_1^2 = x_2^2$

If y_1 is related to $y_2, y_1^2 = y_2^2$.

We can deduce that $(x_1y_1)^2 = (x_2y_2)^2$, and so (x_1y_1) is related to (x_2y_2).

(ii) (a) Yes.
(b) Yes.
If

$$f(0) = g(0) \quad \text{and} \quad h(0) = k(0),$$

then

$$f(0) + h(0) = g(0) + k(0),$$

$$f(0) \times h(0) = g(0) \times k(0).$$

(iii) (a) Yes. If $f' = g'$ and $h' = k'$, then

$$(f + h)' = f' + h' \quad \text{and} \quad (g + k)' = g' + k',$$

and so

$$(f + h)' = (g + k)'.$$

(b) No. For example, if

$$f: x \longmapsto x$$

$$g: x \longmapsto x + 2$$

$$h: x \longmapsto x$$

$$k: x \longmapsto x,$$

then

$$f' = g' \quad \text{and} \quad h' = k'.$$

But

$$f \times h: x \longmapsto x^2$$

and

$$g \times k: x \longmapsto x^2 + 2x$$

and

$$(f \times h)' \neq (g \times k)'.$$ ■

Discussion
* *

We could, of course, start with a function and define the corresponding *natural equivalence relation* (see page 26). If the function is a *morphism*, say, $(A, \circ) \longmapsto (B, \square)$, then the natural equivalence relation (on A) and the operation $\circ$ will be compatible. Let us look at an example to illustrate just what we mean.

Example 3

Example 3

Consider the set of differentiable real functions, and the function on that set

$$D: f \longmapsto f'.$$

We know from *Unit 12, Differentiation I* that D is a morphism for the binary operation of addition in the domain and codomain, and, since D

is many-one, it is a *homomorphism*. We have the following commutative diagram, starting with two elements f, g, of our set:

$$\begin{array}{ccc} (f, g) & \xrightarrow{+} & f + g \\ D\big\downarrow & & \big\downarrow D \\ (Df, Dg) & \xrightarrow{+} & Df + Dg \\ & & = D(f + g) \end{array}$$

Now, for example,

$$x \longmapsto x^2,\ x \longmapsto x^2 - 5,\ x \longmapsto x^2 + 2, \ldots (x \in R)$$

all map to $x \longmapsto 2x$, $x \in R$, and

$$x \longmapsto 3x^3,\ x \longmapsto 3x^3 - 1,\ x \longmapsto 3x^3 + 7, \ldots (x \in R)$$

all map to $x \longmapsto 9x^2$ $(x \in R)$.

So our *natural equivalence relation* sorts

$$x \longmapsto x^2,\ x \longmapsto x^2 - 5,\ x \longmapsto x^2 + 2, \ldots (x \in R)$$

into one equivalence class because they all have the same image, $x \longmapsto 2x$ $(x \in R)$.

Similarly,

$$x \longmapsto 3x^3,\ x \longmapsto 3x^3 - 1,\ x \longmapsto 3x^3 + 7, \ldots (x \in R)$$

all go into another equivalence class because they have the same image $x \longmapsto 9x^2$ $(x \in R)$.

Since the function D is a morphism for addition, the binary operation $+$ and the natural equivalence relation

$$f \rho g \text{ if and only if } Df = Dg$$

are compatible, just as $+$ and D are compatible. Thus if we take any two natural equivalence classes and combine any element of one with any element of the other, we shall always finish with an element of some specific class. For example, the functions

$$(x \longmapsto 3x^3 + x^2) = (x \longmapsto 3x^3) + (x \longmapsto x^2)(x \in R)$$

$$(x \longmapsto 3x^3 + x^2 - 5) = (x \longmapsto 3x^3) + (x \longmapsto x^2 - 5)\,(x \in R)$$

$$(x \longmapsto 3x^3 - 1 + x^2 - 5)$$
$$= (x \longmapsto 3x^3 - 1) + (x \longmapsto x^2 - 5)\,(x \in R)$$

all belong to the equivalence class of elements which map under D to $x \longmapsto 9x^2 + 2x$ $(x \in R)$. ■

19.2.4 Summary

19.2.4

Summary

In this section we began by demonstrating that when we partition a set A we define an equivalence relation on $A: x \rho y$ if x and y belong to the same subset of the partition. Conversely, an equivalence relation partitions a set into non-overlapping subsets, which we call equivalence classes.

We then defined the natural mapping as the function which maps each element to its corresponding equivalence class.

Finally, we considered a binary operation on the set A and its compatibility with the natural mapping, and we saw that when the binary operation and the natural mapping are compatible, then we can *extend* the binary operation to combine equivalence classes.

19.3 ORDER RELATIONS

19.3

19.3.0 Introduction

19.3.0

Introduction

When we considered how a librarian might sort and order the books in his library, we decided to simplify the problem by excluding the possibility of duplicate copies. Once we had decided on an agreed overall order of the general classes (Fiction, Non-fiction and Reference) into which the books were sorted, and on an agreed order within each class, we could then pick up any two of the books in the library and determine which "came first" on the shelves. However, without an agreed ordering of the classes, it was possible to pick up two books at random and not find the ordering question meaningful, as in the case of *Oliver Twist* and *A History of Europe*. There is thus a difference between the two situations, which we should expect to see reflected in the corresponding mathematical models. This difference is between *total ordering* and *partial ordering*.

We can think, if we like, of total ordering as being a stronger property than partial ordering because it orders the whole set. Partial ordering is weaker in the sense that it orders elements within one or more subsets of the set (which need not be disjoint, as they are in the library example). In the library example, when the three general classes themselves are not ordered, these subsets are the classes: Fiction, Non-fiction, Reference. Within each subset we have a defined order for any pair of books, but there is no such order for two books from different classes. Any total ordering is also a partial ordering and so we shall, in general, investigate partial ordering, which we shall often refer to just as *ordering*.

We saw earlier that the reflexive, anti-symmetric and transitive properties are the properties which enable us to order a set; we shall discuss these properties in this section.

We remind you that a relation ρ on a set S is
reflexive if $a \rho a$ for all $a \in S$,
anti-symmetric if whenever $a \rho b$ and $b \rho a$ then $a = b$
and *transitive* if whenever $a \rho b$ and $b \rho c$ then $a \rho c$.

The most familiar relation having these properties is the inequality relation $\leqslant$ on a set of real numbers.

19.3.1 Ordering

19.3.1

We begin by looking at some ordering relations.

Example 1

Example 1

Consider the set of all the subsets of a given set, V, and the inclusion relation on V expressed in terms of $\subseteq$, i.e. $A \subseteq B$ means that A is a subset of B. We have:

$A \subseteq A$ for all $A \in V$ (reflexive)
$A = B$ whenever $A \subseteq B$ and $B \subseteq A$ (anti-symmetric)
$A \subseteq C$ whenever $A \subseteq B$ and $B \subseteq C$ (transitive)

for $A, B, C \in V$. ■

Example 2

Example 2

Consider the set, H, of all human beings, living or dead, and the relationship

$x \rho y$ if x and y are the same person or if x is a direct descendant of y.

We have:

$x \rho x$ for all $x \in H$ (reflexive)
$x = y$ whenever $x \rho y$ and $y \rho x$ (anti-symmetric)
$x \rho z$ whenever $x \rho y$ and $y \rho z$ (transitive)

for $x, y, z \in H$. ■

In both these examples, the relation has the reflexive, anti-symmetric and transitive properties. A relation on a set A which is reflexive, anti-symmetric and transitive is a partial ordering relation on A.

Definition 1
* * *

A partial ordering relation ρ on a set A such that for all $x, y \in A$, $x \neq y$, either $x \rho y$ or $y \rho x$ is called a total ordering relation on A.

In accordance with common terminology, we shall adopt the symbol $\preccurlyeq$ in place of ρ when ρ is an order relation. Similarly, we shall use the symbol $\prec$ to denote the relation $x \prec y$ if $x \preccurlyeq y$ and $x \neq y$.

Notation 1
* * *

When a set has an order relation defined on it, we call it an ordered set.

Definition 2
* * *

We can represent ordered sets by diagrams known as Hasse diagrams, some examples of which are given below.

Definition 3
* * *

Example 3

Example 3

This could represent the set of integers $\{1, 2, 4, 8\}$ together with $\preccurlyeq$ interpreted as "is a factor of". ■

Example 4

Example 4

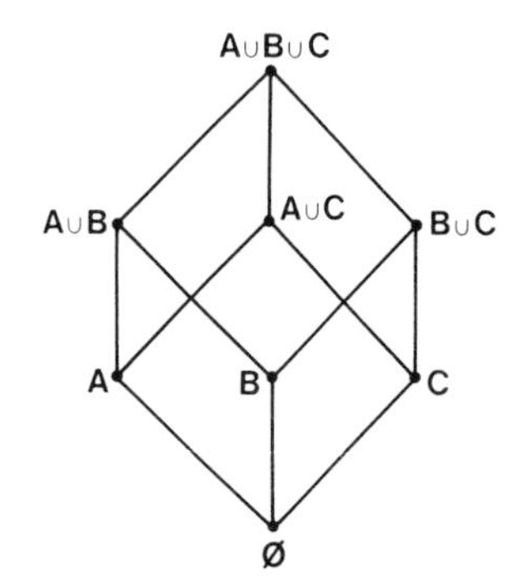

This could represent disjoint sets, $\varnothing$, A, B, C, combined under the binary operation of union (as shown) to form the set $\{\varnothing, A, B, C, A \cup B, A \cup C, B \cup C, A \cup B \cup C\}$, with $\preccurlyeq$ interpreted as "is a subset of".

(Remember that we consider the empty set $\varnothing$ to be a subset of every set.) ■

Discussion
* *

These diagrams are an attempt to show pictorially some of the ideas we have been discussing. The ordering of the sets is illustrated by the vertical status of the elements; thus, in Example 4, the subset $\{A \cup B \cup C, \varnothing, B, B \cup C\}$ is ordered into $\varnothing - B - B \cup C - A \cup B \cup C$. The subset $\{A \cup B, B \cup C\}$ is not ordered.

We can see from the diagram in Example 3 that the whole set is ordered—there is a link between every pair of elements. On the other hand, Example 4 is an example of *partial* ordering—there is not a link between every pair of elements. Notice that we do not have to put in *all* the direct links (i.e. Hasse diagrams are not the same as the diagrams we encountered previously in this text, where we joined each pair of related elements). We can see that A is a subset of $A \cup B \cup C$ because it is linked via $A \cup B$.

Exercise 1

Exercise 1
(3 minutes)

Check that the following cases give examples of ordered sets, and draw the corresponding Hasse diagrams.

(i) The set {Oxford, Birmingham, London, Manchester, Exeter, Glasgow}, with the relation "does not lie to the south of".

(ii)

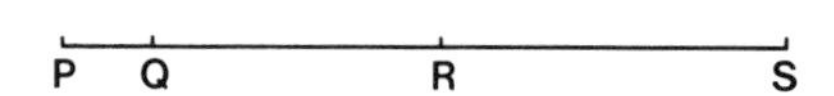

The set of pairs of points from the set $\{P, Q, R, S\}$ of four points on a straight line as in the diagram, with the relation $(x, y) \preccurlyeq (w, z)$ if the interval xy is contained in or equal to the interval wz, where x, y, w, z represent P, Q, R, S in some order.

For example, the interval QR is inside the interval PS, so that $(Q, R) \preccurlyeq (P, S)$; but $(P, R) \not\preccurlyeq (Q, R)$, because the interval PR is not included in the interval QR. ■

Discussion
* *

In *Unit 2* we defined the *absolute error bound* to be the maximum possible value of the magnitude of the absolute error. That is, if x is a number which is an estimate of the exact number X, and the absolute error bound is ε, then we have

$$|X - x| \leqslant \varepsilon$$

i.e.

$$-\varepsilon \leqslant X - x \leqslant \varepsilon$$

so

$$x - \varepsilon \leqslant X \leqslant x + \varepsilon$$

We also used the concept of a bound when discussing the general Taylor approximation in *Unit 14*.

Main Text
* * *

We can now extend the definition of a *bound* and define bounds of a subset of a partially or totally ordered set.

Definition 4
* * *

An upper bound of a subset S of an ordered set P is any element u of P for which $a \preccurlyeq u$ for all elements $a \in S$.

We can define a *lower bound* similarly.

Definition 5
* * *

A lower bound of a subset S of an ordered set P is any element l of P for which $l \preccurlyeq a$ for all elements $a \in S$.

Notice that the upper and lower bounds must belong to the ordered set P under consideration, but *not necessarily* to the subset S in question.

Upper and lower bounds need not be unique. Thus, referring again to the diagram of Example 4, and considering the subset $\{\varnothing, B, C\}$, we see that both $B \cup C$ and $A \cup B \cup C$ are upper bounds of this subset. On the other hand, when we look for lower bounds of the same subset, we find that there is only one, namely $\varnothing$.

This happens to be the case here, because $\varnothing$ is the "lowest" element of P. Were there any element "below" $\varnothing$, this would also be a lower bound of $\{\varnothing, B, C\}$. For example, the subset $\{A \cup B, B \cup C, B\}$, has lower bounds B and $\varnothing$.

Example 5

Example 5

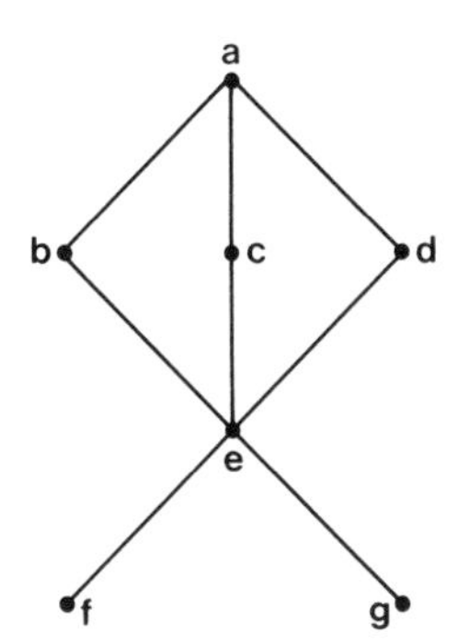

Here, for the subset $\{b, c, d, e\}$, a is the only upper bound, but each of e, f, g is a lower bound of the subset. If, however, we consider the set of all lower bounds of $\{b, c, d, e\}$, namely $\{e, f, g\}$, we see that one element is "higher" in the diagram than the rest, in this case e.

We call e the *greatest lower bound* of $\{b, c, d, e\}$. ■

Definition 6
* * *

An element $l_g \in P$ is the greatest lower bound of a subset S of an ordered set P, if l_g is a lower bound of S and $l \preccurlyeq l_g$ for every lower bound l of S.

(continued on page 36)

Solution 1

Solution 1

(i)

(ii)

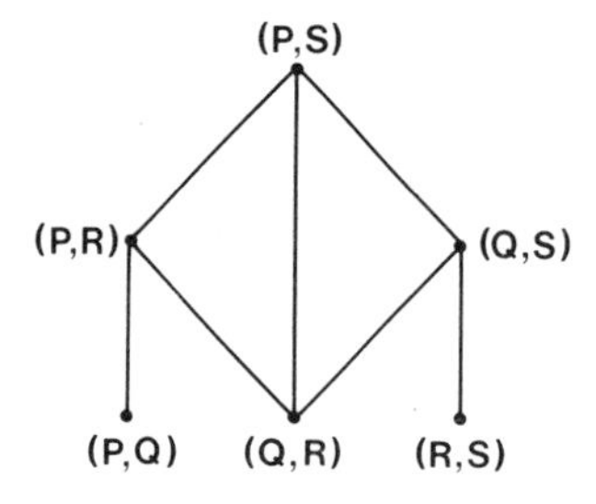

This is an example of *partial* ordering, as opposed to *total* ordering, because there is no link, for instance, between (P, R) and (Q, S). ■

(*continued from page 35*)

In a similar manner, we can define the *least upper bound* of a subset of an ordered set.

Definition 7
* * *

An element $u_g \in P$ is the least upper bound of a subset S of an ordered set P, if u_g is an upper bound of S and $u_g \preccurlyeq u$ for every upper bound u of S.

Looking back to Example 5, we see, for example, that the least upper bound of $\{b, c, e, f\}$ is a, the least upper bound of $\{b, e, f\}$ is b, and the greatest lower bound of $\{a, b, c, d\}$ is e.

It is important to note that the least upper bound and greatest lower bound *may not exist* for some subsets of a given partially ordered set. For example, $\{e, f, g\}$ has no greatest lower bound for the set of Example 5. It has two lower bounds, f and g, but we have neither $f \preccurlyeq g$ nor $g \preccurlyeq f$. When they exist however, both the greatest lower bound and the least upper bound of a given subset are *unique*. (See Exercise 2.)

Exercise 2

Exercise 2
(3 minutes)

(i) Consider the set $\{1, 1 + \frac{1}{2}, 1 + \frac{1}{2} + \frac{1}{4}, 1 + \frac{1}{2} + \frac{1}{4} + \frac{1}{8}, \ldots\}$ considered as a subset of the reals with the relation $\leqslant$ as $\preccurlyeq$. What is the greatest lower bound and what is the least upper bound?

(ii) By writing down the missing symbols and words, complete the following proof that, when it exists, the greatest lower bound is unique.
Suppose u_1 and u_2 are two greatest lower bounds and $u_1 \neq u_2$. Then u_2 is necessarily a lower bound, and so, since u_1 is a *greatest* lower bound,

$$u_2 \; \boxed{} \; u_1. \qquad (1)$$

Also u_1 is necessarily a lower bound, and so, since u_2 is a *greatest* lower bound,

$$u_1 \; \boxed{} \; u_2. \qquad (2)$$

Since $\preccurlyeq$ is a partial ordering relation, statements (1) and (2) together imply that

$$u_1 \; \boxed{} \; u_2$$

which contradicts our original assumption. This is an example of

proof by $\boxed{}$. ■

We shall end this sub-section with a further example.

Example 6

Example 6

Consider the general Venn diagram for two sets A and B:

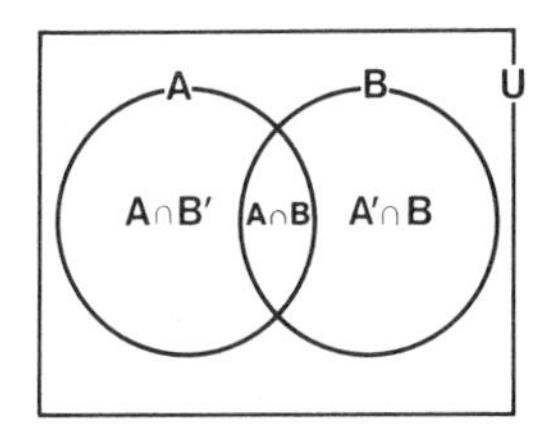

We can obtain an interesting set of sets by starting with the empty set, $\varnothing$, then taking the four basic regions of the Venn diagram one at a time, two at a time, three at a time, and finally four at a time (giving us the universal set U). We obtain the set of subsets:

$\{\varnothing,$
$A' \cap B', A \cap B', A' \cap B, A \cap B,$
$B', A', (A' \cap B') \cup (A \cap B), (A' \cap B) \cup (A \cap B'), A, B,$
$A' \cup B', A \cup B', A' \cup B, A \cup B,$
$U\}$

For the relation of set inclusion, $\subseteq$, the Hasse diagram is:

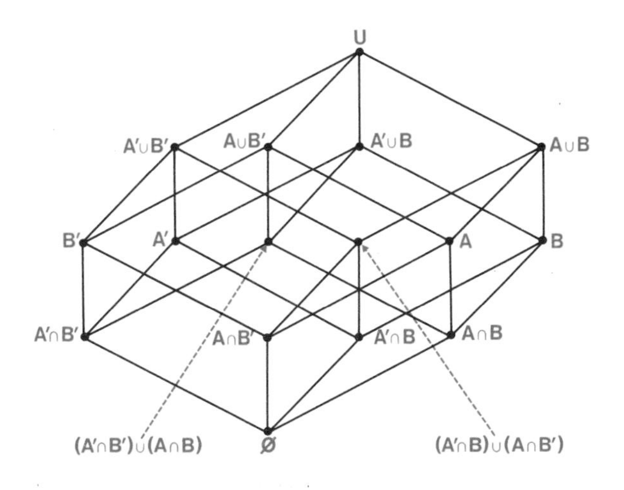

(*continued on page 38*)

Solution 2

(i) The greatest lower bound is 1. The least upper bound is 2.

(ii)

$\preccurlyeq$

$\preccurlyeq$

=

contradiction

■

(*continued from page 37*)

(You may be interested in the fact that this is what a "four-dimensional cube" would look like projected on to a plane, just as Example 4 shows such a projection of an ordinary three-dimensional cube.)

Let us now look for greatest lower bounds and least upper bounds for some sets of subsets. We have, for example,

g.l.b. $\{A' \cap B', A \cap B', A' \cap B\} = \varnothing$
g.l.b. $\{A \cup B', A' \cup B\} = (A' \cap B') \cup (A \cap B)$
l.u.b. $\{\varnothing, A' \cap B', A \cap B', B'\} = B'$
l.u.b. $\{A, B, A'\} = U$
... etc.,

where g.l.b. and l.u.b. denote greatest lower bound and least upper bound respectively.

Notation 2
* * *

If you look at these examples again, you will see that the g.l.b. is simply the intersection of all the sets in the chosen set of subsets, and the l.u.b. is simply the union of all the sets in the chosen set of subsets. ■

In Example 6, we have shown that a very important connection exists between {l.u.b., g.l.b.} and {union, intersection}. In this particular example, you may have realized that we would necessarily have been able to determine the least upper bound and the greatest lower bound no matter what subset of the ordered set we chose to investigate.

An ordered set like this, for which *every* subset has a g.l.b. and l.u.b., is called a lattice, and the theory of lattices is itself a topic in its own right within mathematics.

Definition 8
*

19.3.2 Summary

19.3.2

We summarize what we have covered in this section.

First, we defined the term *ordering* and gave a useful way of representing such a relation on a set—the *Hasse diagram.* We then defined a *total ordering relation* and a *partial ordering relation.* We defined a *lower bound* and an *upper bound* of a subset of an ordered set, and found that a subset may have more than one of each of these. It may have a *greatest lower bound* and a *least upper bound,* and if it does, each is *unique.*

We have come a long way from our young child sorting and ordering his cubes, and from our librarian organizing the books in his library.

There are many subjects belonging to the more recently discovered areas of mathematics, which have their beginnings in the work we have considered in this text.

Unit No.		Title of Text
1		Functions
2		Errors and Accuracy
3		Operations and Morphisms
4		Finite Differences
5	NO TEXT	
6		Inequalities
7		Sequences and Limits I
8		Computing I
9		Integration I
10	NO TEXT	
11		Logic I — Boolean Algebra
12		Differentiation I
13		Integration II
14		Sequences and Limits II
15		Differentiation II
16		Probability and Statistics I
17		Logic II — Proof
18		Probability and Statistics II
19		Relations
20		Computing II
21		Probability and Statistics III
22		Linear Algebra I
23		Linear Algebra II
24		Differential Equations I
25	NO TEXT	
26		Linear Algebra III
27		Complex Numbers I
28		Linear Algebra IV
29		Complex Numbers II
30		Groups I
31		Differential Equations II
32	NO TEXT	
33		Groups II
34		Number Systems
35		Topology
36		Mathematical Structures